Accelerates Academic Langu

OXFORD
ILLUSTRATED
MATH
DICTIONARY

OXFORD
UNIVERSITY PRESS

OXFORD
UNIVERSITY PRESS

198 Madison Avenue
New York, NY 10016 USA

Great Clarendon Street, Oxford, OX2 6DP,
United Kingdom

Oxford University Press is a department of the University of
Oxford. It furthers the University's objective of excellence in research,
scholarship, and education by publishing worldwide. Oxford is a
registered trade mark of Oxford University Press in the UK and in
certain other countries

© Oxford University Press 2012

Library of Congress Cataloging-in-Publication Data

The Oxford illustrated math dictionary.
    p. cm.
    1. Mathematics–Dictionaries. 2. Picture dictionaries, English.
    I. Title: Illustrated math dictionary.
    QA5.O94 2012
    510.3–dc23
    2011033043

The moral rights of the author have been asserted

First published in 2012
5 6 7 8 9 10   17

General Manager, American ELT: Laura Pearson
Publisher: Stephanie Karras
Development Editor: Brandon Lord
Director, ADP: Susan Sanguily
Design Manager: Lisa Donovan
Cover Design: Yin Ling Wong
Electronic Production Manager: Julie Armstrong
Production Artist: Elissa Santos
Image Manager: Trisha Masterson
Image Editor: Liaht Pashayan
Senior Controller, Manufacturing: Eve Wong

ISBN: 978 0 19 407128 4

Printed in China

This book is printed on paper from certified and well-managed
sources

ACKNOWLEDGEMENTS

Illustrations by: Argosy, Fitz Hammond, Steve May, Colin Mier, Mark
Ruffle, John Walker

*The publishers would like to thank the following for their kind permission to
reproduce photographs:*

Cover photos: Art Stock Photos/Alamy (Origami); Les Cunliffe/
agefotostock (calculator); Lew Robertson/Corbis (compass); Florian
Augustin/shutterstock.com (equations).

Pg. 2 age fotostock/SuperStock; pg. 5 PhotoTalk/istockphoto.com; pg.
Leslie Garland Picture Library/Alamy; pg. 13, 15 NorGal/shutterstock.
com; pg. 17 magicoven/shutterstock.com (penny, dime); pg. 17 Getty
Images (nickel); pg. 20 PhotoTalk/istockphoto.com (digital clock); pg.
20 Edyta Pawlowska/shutterstock (analog clock); pg. 21 Getty Images;
pg. 23 Alexander Shirokov/istockphoto.com (drawing compass); pg. 23
Dmitry Rukhlenko/shutterstock.com (navigating compass); pg. 24, 27
Brenda A. Carson/istockphoto.com; pg. 29 Golden Pixels LLC/Alamy;
pg. 39 Lana Langlois/shutterstock.com; pg. 41 PhotoTalk/istockphoto.
com; pg. 51 mattesimages/shutterstock.com (heads/tails); pg. 51
Dennis Kitchen Studio, Inc./Oxford University Press (dollar); pg. 51 ATV
Studio/shutterstock.com (quarter); pg. 51 Getty Images (nickel); pg.
51 magicoven/shutterstock.com (dime); pg. 60 Keith Leighton/Alamy
(raisin); pg. 60 Corbis/Oxford University Press (paper clip); pg. 62 Alina
Solovyova-Vincent/istockphoto.com; pg. 74 craftvision/istockphoto.
com; pg. 79 foodanddrinkphotos co/age fotostock; pg. 80 Leslie Garland
Picture Library/Alamy; pg. 89 Rick Fischer/Masterfile; pg. 96 Photodisc/
Oxford University Press; pg. 98 kak2s/shutterstock.com; pg. 108
mattesimages/shutterstock.com; pg. 120 Elena Leonova/shutterstock.
com (flowers); pg. 120 Susan Kopecky/istockphoto.com (ruler); pg.
121 Don Farrall/Getty Images (scale); pg. 121 rzelich/istockphoto.com
(thermometer); pg. 127 Ingram/Oxford University Press (basketball); pg.
127 Lisa F. Young/shutterstock.com (football); pg. 135 Thomas Northcut/
Getty Images; pg. 137 PhotoTalk/istockphoto.com (digital clock); pg. 1
Thomas Northcut/Getty Images (stopwatch); pg. 138 Maksim Toome/
shutterstock.com; pg. 147 Don Farrall/Getty Images.

Contents

Acknowledgments

Our National Standards and Mathematics Consultant

Vena M. Long, Ed.D.
Dr. Long is a professor of mathematics education and an associate dean for research and professional development at the University of Tennessee. She served on the board of directors of the National Council of Teachers of Mathematics from 2006 to 2010 and is also active in local- and state-level professional organizations.

The publisher would like to acknowledge the following individuals for their invaluable feedback during the development of this program.

Judy Dean: Austin Independent School District, Austin, TX

Dr. Sandra Stockdale: Collier County Public Schools, Naples, FL

Noreen N. Kraebel: Fox Hollow Elementary, Port Richey, FL

Iryna Khits: Hopewell High School, Charlotte, NC

Anastasia Babayan: Huntington Beach, CA

LaTisha Ford, Christa Tillman-Young: James Coble Middle School, Arlington, TX

Tracy Thompson: Lowrey Middle School, Dearborn, MI

Tamara Lopez: McCoy Elementary, Orlando, FL

Janet E. Lasky: Montgomery County Public Schools, Rockville, MD

Susan Welch: District School Board of Pasco County, Land O'Lakes, FL

Mamiko Nakata: Prince George's County Public Schools, Silver Spring, MD

Ranada Young: Roberts High School, Salem, OR

Weston Calbreath: Steele Creek Elementary, Charlotte, NC

Nehmat Sabra: Stout Middle School, Dearborn, MI

Anne Hagerman Wilcox, Lisa Nelson: Wendell School District, Wendell, ID

Each part of the entry helps you learn the term.

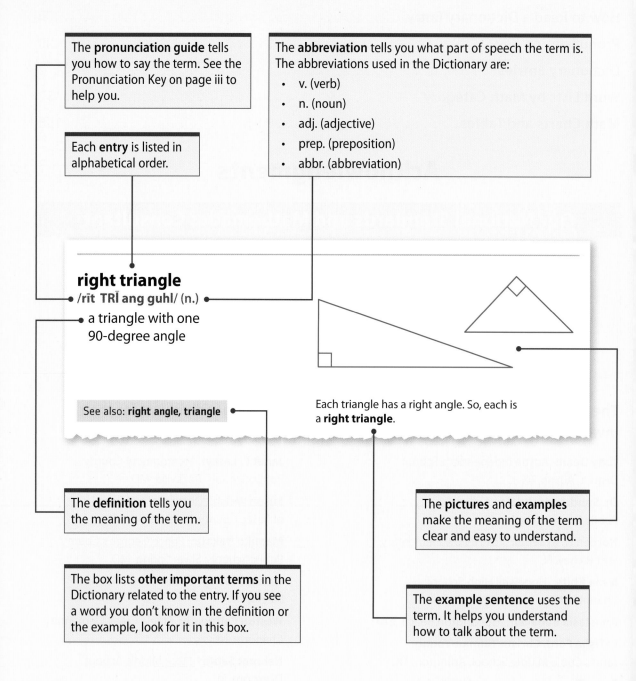

The **pronunciation guide** tells you how to say the term. See the Pronunciation Key on page iii to help you.

The **abbreviation** tells you what part of speech the term is. The abbreviations used in the Dictionary are:

- v. (verb)
- n. (noun)
- adj. (adjective)
- prep. (preposition)
- abbr. (abbreviation)

Each **entry** is listed in alphabetical order.

right triangle
/rīt TRĪ ang guhl/ (n.)

a triangle with one 90-degree angle

See also: **right angle, triangle**

Each triangle has a right angle. So, each is a **right triangle**.

The **definition** tells you the meaning of the term.

The **pictures** and **examples** make the meaning of the term clear and easy to understand.

The box lists **other important terms** in the Dictionary related to the entry. If you see a word you don't know in the definition or the example, look for it in this box.

The **example sentence** uses the term. It helps you understand how to talk about the term.

Pronunciation Key

Vowel Sounds	Like the Sound in . . .	Consonant Sounds	Like the Sound in . . .
a	bat, map	b	boy, job
ay	ate, say	ch	chair, lunch
ah	father, calm	d	day, mud
air	care, fair	f	fall, brief
ar	car, far	g	gone, bug
e	met, step	h	hear, hail
ee	me, equal	j	jaw, enjoy, gel
ur	fern, stir, burn	k	key, cold, took, track
i	if, fit	l	lake, tool
ī	ice, time, fly	m	my, jam
o	stop, clock	n	night, run
oh	ocean, load	ng	song, bring
or	orange, orbit	p	pay, stop
aw	jaw, talk	r	rake, press
oi	soil, boy	s	slow, bus
ow	out, flower	sh	short, bush
uh	cut, summer	t	tip, out
u	full, put	th	thick, bath
oo	soon, prove	TH	there, weather
		v	voice, save
		w	won, winter
		y	yes, young
		z	zoo, freeze
		zh	treasure

Stressed syllables appear in capital letters: /suh LOO shuhn/

iii

abacus /A buh kuhs/ (n.)

a tool used to help make mental calculations

See also: **calculation, counter, place value**

An **abacus** uses counters to show place value.

accurate /A kyuh ruht/ (adj.)

correct; exactly right without any mistakes

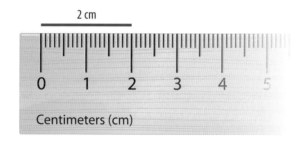

See also: **measurement**

An **accurate** measurement is 2 cm.

acute angle
/uh KYOOT ANG guhl/ (n.)

an angle that measures less than 90°

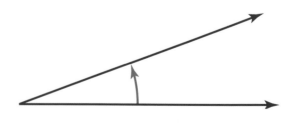

See also: **angle, degree, obtuse angle, right angle**

A 20° angle is an **acute angle**.

acute triangle
/uh KYOOT TRĪ ang guhl/ (n.)

a triangle with each angle less than 90°

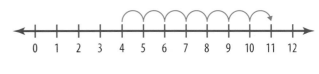

scalene triangle

equilateral triangle

isosceles triangle

See also: **equilateral triangle, isosceles triangle, obtuse triangle, scalene triangle**

An **acute triangle** can be scalene. It can also be equilateral or isosceles.

add (+) /ad/ (v.)

to combine numbers to make a new number

$4 + 7 = 11$

Add four and seven to make eleven.

The sign for **add** is +. It is called the plus sign.

See also: **operation, plus, sum, total**

addition (+) /uh DISH uhn/ (n.)

the act of combining numbers to make a new number

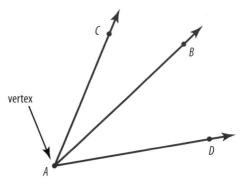

$4 + 7 = 11$

See also: **operation, plus, sum, total**

You can use a number line to show **addition**.

adjacent /uh JAY suhnt/ (adj.)

next to, side by side

vertex

∠CAB is adjacent to ∠BAD.

See also: **angle, ray, vertex**

Adjacent angles share a side and vertex.

3

after /AF tur/ (prep.)
next in order

June 2nd is **after** June 1st.
The number 51 is **after** the number 50.
Lunch is **after** breakfast.

See also: **before, order**

---

afternoon /af tur NOON/ (n.)
the time of day between noon
and evening

School usually ends in the **afternoon**.

See also: **day, morning, night**

---

algebra /AL juh bruh/ (n.)
math about the rules of
operations and relations
of numbers

Use **algebra** to find unknown values in equations.
Chris is four years older than his brother, Joe.
Let C = Chris's age. Let J = Joe's age.
Then, $C = 4 + J$. When Joe is 10, Chris is 14.

See also: **equation, symbol, variable**

---

altitude /AL tuh tood/ (n.)
a line segment that joins
a figure's vertex to its opposite
side; it is also perpendicular
to the opposite side

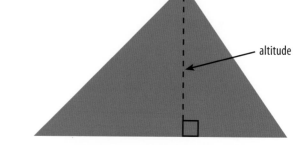

See also: **base, height, right angle, vertex**

The dashed line is an **altitude** of the triangle.
The **altitude** and the base form a right angle.

altogether
/awl tuh GE THur/ (adv.)
in total; in all

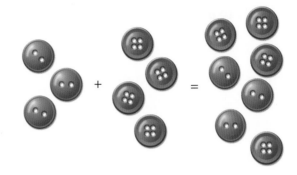

There are seven buttons **altogether**.

See also: **sum, total**

---

a.m. /ay em/ (abbr.)
before noon; morning

I like to have breakfast at 7:00 **a.m.**

See also: **morning, night, p.m.**

---

angle (∠) /ANG guhl/ (n.)
two rays that share an endpoint

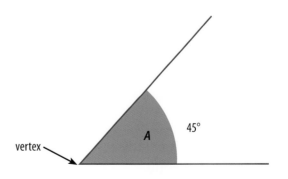

vertex

A

45°

This **angle** is called ∠A.
∠A measures 45°.

See also: **degree, endpoint, ray**

answer /AN sur/ (n.)

    1. a reply to a question

> My **answer** to your question is "yes."

    2. the result of solving a problem

> 20 − 9 = ?
>
> The **answer** is 11.

See also: **solution**

---

apex /AY peks/ (n.)

    the vertex farthest away from
the base of a shape

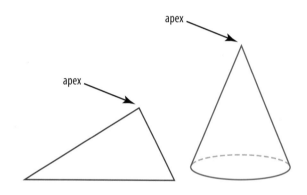

See also: **base, solid, vertex**

A triangle can have an **apex**. A solid can have an **apex**.

---

approximate /uh PRAHK suh mayt/ (v.)

    to make an estimate that is close
to the exact answer

> I **approximate** the cost
> of these groceries is $50.

See also: **estimate, round**

---

approximately
/uh PRAHK suh muht lee/ (adv.)

    close to, but not exactly

> π = 3.141592...
>
> The value of pi is **approximately** 3.14.

See also: **estimate, estimation, round**

arc /ark/ (n.)

part of a circle or any curve

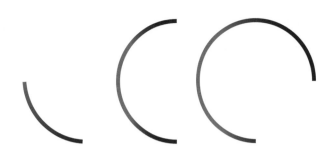

See also: **circle, compass, curve**

You can draw an **arc** with a compass.

---

area /AIR ee uh/ (n.)

the amount of surface
inside a shape

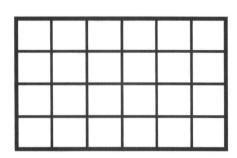

See also: **perimeter, square, surface**

We measure **area** in square units. This rectangle has
an **area** of 24 square units.

---

arithmetic /uh RITH muh tik/ (n.)

counting and calculating
with numbers

She is good at **arithmetic**. She can add,
subtract, divide, and multiply in her head.

See also: **calculate, count**

---

arrange /uh RAYNJ/ (v.)

to put things in a particular order

456, 465; 546, 564; 645, 654

You can **arrange** the numbers 4, 5, and 6
in six ways.

See also: **column, order, row**

array /uh RAY/ (n.)

objects arranged in rows and columns

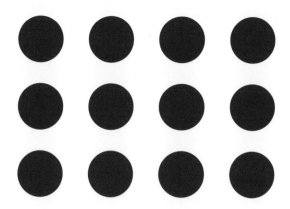

This **array** has 3 rows and 4 columns. It models the multiplication fact 3×4.

See also: **column, model, multiplication, row**

ascending /uh SEN ding/ (adj.)

becoming greater in size

1, 10, 14, 99, 104, 222, 385

These numbers are in **ascending** order.

See also: **descending, order**

Associative Property

/uh SOH shee ay tiv PROP ur tee/ (n.)

the rule that says you can add or multiply numbers in any order

$$(2 \times 3) \times 4 = 2 \times (3 \times 4)$$
$$5 + (6 + 7) = (5 + 6) + 7$$

The **Associative Property** makes arithmetic easier.

See also: **group, order, product, sum**

attribute /A truh byoot/ (n.)

a feature of an object or a set

$$\{2, 4, 6, 8, 10\}$$

All the numbers in the set have the **attribute** of being even.

See also: **right angle, set, side, square**

average /A vur ij/ (n.)
the mean of a set of numbers

See also: **mean, median, mode, set**

3   3   4   6   9

1. Find the sum of the numbers in the set.

$3 + 3 + 4 + 6 + 9 = 25$

2. Divide the sum by the total number of numbers.

$25 \div 5 = 5$

The **average** is 5.

---

axis /AK suhs/ (n.)
1. a number line in a coordinate grid

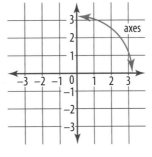

The horizontal **axis** goes from −3 to 3. The vertical **axis** goes from −3 to 3.

2. a line that divides a figure in two equal halves

See also: **coordinate grid, line of symmetry, x-axis, y-axis**

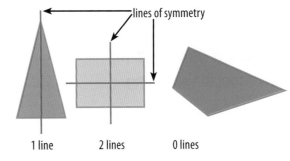

1 line   2 lines   0 lines

This figure's **axis** is the same as its line of symmetry.

---

axis of rotation
/AK suhs uhv roh TAY shuhn/ (n.)

a straight line through the center of a solid shape

See also: **rotate, solid, sphere**

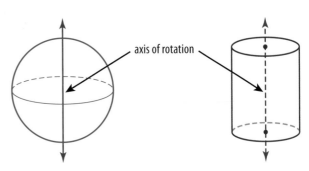

This sphere rotates on its **axis of rotation**.

Bb

backward /BAK wurd/ (adv.)

in reverse; in a way or direction that is the opposite of the usual

10, 9, 8, 7, 6, 5, 4, 3, 2, 1

You can count **backward** from 10. You begin at 10. You end at 1.

See also: **count, forward**

---

balance /BA luhns/ (v.)

to keep one side of an equation equal to the other side

$$3 + b = 12$$
$$3 - 3 + b = 12 - 3$$
$$b = 9$$

Balance the equation to find b. Subtract 3 from both sides.

See also: **equation, solve, variable**

---

balance /BA luhns/ (n.)

1. a tool for weighing

Use the **balance** to weigh materials.

2. the amount of money in a bank account

I put $100 in my bank account. I took out $25. The **balance** of my account is now $75.

See also: **interest, measure, scale, weight**

bar graph /bar graf/ (n.)

a picture with rectangles that shows data

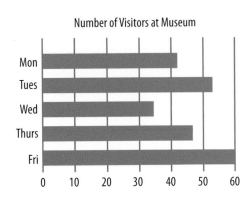

Number of Visitors at Museum

See also: **data, graph, horizontal, rectangle**

The **bar graph** shows the museum had the most visitors on Friday.

base /bays/ (n.)

1. the part a figure stands on; usually its bottom

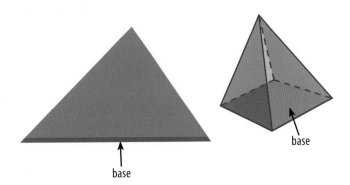

base

base

A triangle and a cone each have a **base**. Any flat side of a shape or solid can be a **base**.

2. the number of digits in a number system

Hundreds	Tens	Ones
10 · 10	10	1

The United States uses a **base**-10 number system. (0, 1, 2, 3, 4, 5, 6, 7, 8, 9)

3. a number that is raised to an exponent

$8^3 = 512$

The number 8 is the **base** raised to the power of 3.

See also: **digit, exponent, figure, power**

before /bee FOR/ (prep.)
 coming in front of something
 in order

| August 14th is **before** August 15th.
The number 22 is **before** 23.
Breakfast comes **before** lunch. |

See also: **after, order**

---

big /**big**/ (adj.)
 large in size

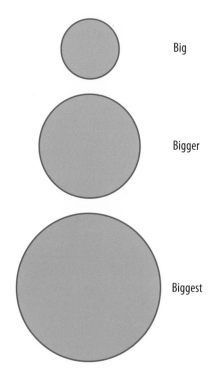

Big

Bigger

Biggest

See also: **size**

How **big** is each circle? Each is bigger or smaller than
the one next to it.

---

billion /**BIL yuhn**/ (n.)
 a number that is one thousand
 times one million

The number 1 **billion** has nine zeros.
It can be written in different ways.

$$1{,}000{,}000{,}000 = 10^9$$

See also: **million, times, zero**

bisect /BĪ sekt/ (v.)

to divide into two equal parts

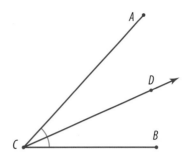

Draw a ray *CD* to **bisect** angle *ACB*. Angle *ACD* is equal to angle *DCB*.

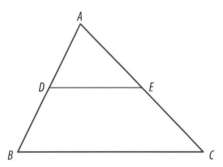

See also: **angle, congruent, line segment**

$\overline{DE}$ **bisects** $\overline{AB}$ and $\overline{AC}$. $\overline{AD}$ equals $\overline{DB}$, and $\overline{AE}$ equals $\overline{EC}$.

---

brackets ([ ]) /BRA kuhts/ (n.)

symbols that group certain numbers together

$$[(2 + 3) \times (2 + 1)] \times 2 = 30$$

Solve inside the **brackets** first.
The **brackets** above are red.

See also: **order of operations, parentheses, symbol**

---

button /BUT n/ (n.)

a part of a calculator that you press

Press the "+" **button** to add the numbers.

See also: **calculator**

Cc

C /see/ (n.)
the Roman numeral
representing 100

C = 100
CI = 101
CV = 105
CC = 200

See also: **numeral, Roman numerals**

---

calculate /KAL kyuh layt/ (v.)
to work out the answer
to a number problem

Calculate the answer: $\frac{3}{4}$ of 160

$\frac{3}{4} \times 160$

$\frac{3}{4} \times \frac{160}{1}$

$\frac{480}{4}$

$480 \div 4 = 120$

See also: **arithmetic, operation, solve**

---

calculation /kal kyuh LAY shuhn/ (n.)
the steps to answer an
arithmetic problem

She made mistakes in her **calculation**.
She added the numbers.
She should have multiplied them.

See also: **answer, calculate, solution, solve**

calculator /KAL kyuh lay tur/ (n.)
a small machine that calculates math problems

See also: **arithmetic, calculate, calculation**

Use your **calculator** to do difficult arithmetic.

calendar /KAL uhn dur/ (n.)
a chart that shows the days, weeks, and months of a year

January 2012

SUN	MON	TUE	WED	THU	FRI	SAT
1	2	3	4	5	6	7
8	9	10	11	12	13	14
15	16	17	18	19	20	21
22	23	24	25	26	27	28
29	30	31				

See also: **day, month**

The **calendar** shows the month of January. It shows that January 1st is on a Sunday.

capacity /kuh PAS uh tee/ (n.)
how much something can hold

4 oz
3 ½
3
2 ½
2
1 ½
1
½

See also: **cup, milliliter, volume**

The **capacity** of the measuring cup is 4 ounces. There are 3 ounces of juice in the cup.

Celsius (C) /SEL see uhs/ (n.)
a scale for measuring temperature

Water freezes at 0° **Celsius**.
Water boils at 100° **Celsius**.

See also: **degree, Fahrenheit, scale, temperature**

15

center /SEN tur/ (n.)
exactly in the middle

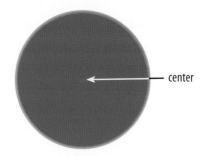
center

See also: **circle, distance, point**

A circle's **center** is the same distance from all points on the circle.

---

center of rotation
/SEN tur uhv roh TAY shuhn/ (n.)
the point a figure rotates or spins around

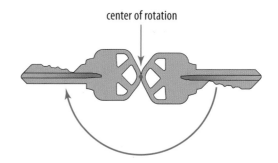
center of rotation

See also: **clockwise, counterclockwise, point, rotate**

The key rotates clockwise around the **center of rotation**.

---

centiliter (cl) /SEN tuh lee tur/ (n.)
a unit of volume equal to one hundredth of a liter

$$1 \text{ centiliter} = \frac{1}{100} \text{ liter}$$

$$100 \text{ centiliters} = 1 \text{ liter}$$

See also: **capacity, liter, metric system, milliliter**

---

centimeter (cm) /SEN tuh mee tur/ (n.)
one hundredth of a meter

$$1 \text{ centimeter} = \frac{1}{100} \text{ meter}$$

$$100 \text{ centimeters} = 1 \text{ meter}$$

See also: **length, meter, metric system, millimeter**

century /SEN chur ee/ (n.)
a period of one hundred years

People didn't have computers a **century** ago.

See also: **millennium**

---

certain /SURT n/ (adj.)
positive; without doubt

See also: **impossible, likely, probability, probable**

This is a six-sided number cube. It is **certain** I will roll a number less than 10. It is not **certain** I will roll a 3.

---

change /chaynj/ (n.)
1. coins

I have some **change** on my desk. There is a penny, a nickel, and a dime.

2. money returned after paying for something

Lunch cost $16.50.
I paid with a $20 bill.
My **change** was $3.50, or $20 – $16.50.

See also: **coin**

chart /chart/ (n.)
something that shows
information in an organized way

Grade	Number of Students
A	IIII
B	HHt I
C	HHt HHt
D	III
F	HHt

See also: **data, graph, pie chart, tally chart**

The tally **chart** shows grades on the last test. It shows four students got As.

---

cheap /cheep/ (adj.)
low in cost or price

The shirt is **cheap**. It costs only $4.
The other shirt is not **cheap**. It costs $35.

See also: **cost, price**

---

check /chek/ (v.)
to make sure an answer
or calculation is correct

I subtract 278 from 500. My answer is 222.
I can **check** my answer by adding 222 and 278.

See also: **answer, calculation, reasonable, solution**

---

choose /chooz/ (v.)
to pick or select from a group

There is jam, banana, and peanut butter.
You can **choose** two for your toast.
How many possible pairs can you **choose**?
Make a list to find the answer:

jam and banana

jam and peanut butter

banana and peanut butter

You can **choose** one of three pairs.

See also: **different, pair**

18

chord /kord/ (n.)

a line segment that joins two points on a circle

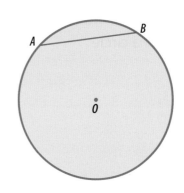

See also: **circle, diameter, line segment, radius**

$\overline{AB}$ is a **chord** of circle O.

circle /SUR kuhl/ (n.)

a figure made up of points the same distance from another point

center ⟶ · ⟵ diameter

⟵ radius

See also: **center, distance, point**

The red curved line represents a **circle**.

circle graph /SUR kuhl graf/ (n.)

a circle divided to represent different data

The Hours in Jack's Day

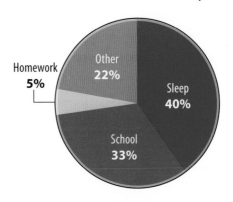

Homework 5% Other 22% Sleep 40% School 33%

A **circle graph** shows data as parts of a whole. This **circle graph** shows the hours of Jack's day. Jack spends 33 percent of his day in school.

See also: **graph, percent, pie chart, sector**

19

circular /SUR kyuh lur/ (adj.)

in the shape of a circle

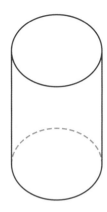

See also: **circle, cylinder, face**

A cylinder has a **circular** top.

circumference
/sur KUM fur uhns/ (n.)

the distance around a circle

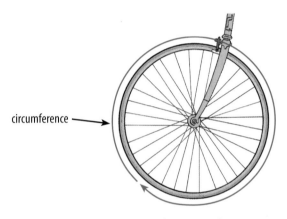

circumference

The formula for the **circumference** of a circle is $C = \pi \times d$. C is the circle's **circumference**. Its diameter is d.

See also: **circle, diameter, distance, pi**

clock /klok/ (n.)

a machine that shows the time

See also: **hour hand, minute hand, time**

Each **clock** shows it is 7:00.

clockwise   /KLOK wīz/ (adv.)

in the same direction that
the hands of a clock turn

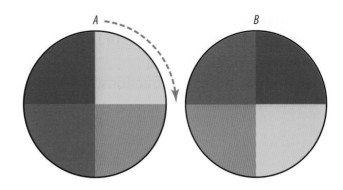

See also: **counterclockwise, minute hand, rotate**

Rotate circle *A* **clockwise** a quarter turn. The result is circle *B*.

coin   /koin/ (n.)

a piece of money usually made
from metal

See also: **change**

The **coin** is a nickel. It is worth five cents.

column   /KOL uhm/ (n.)

numbers or information arranged
in a vertical way

a column

1	2	3	4
5	6	7	8
9	10	11	12
13	14	15	16

See also: **array, row, vertical**

The second **column** of the array is orange. The numbers in the **column** are 2, 6, 10, and 14.

common denominator
/KOM uhn di NOM uh nay tur/ (n.)

a shared multiple of the denominators of two fractions

Find a **common denominator** to add $\frac{2}{3}$ and $\frac{1}{4}$.

1. Rewrite each fraction with a **common denominator**.

$$\frac{2 \times 4}{3 \times 4} + \frac{1 \times 3}{4 \times 3}$$

2. Multiply. The **common denominator** is 12.

$$\frac{8}{12} + \frac{3}{12}$$

3. Add the numerators.

$$\frac{11}{12}$$

See also: **denominator, fraction, multiple, numerator**

Commutative Property
/kuh MYOO tuh tiv PROP ur tee/ (n.)

the rule that says you can add or multiply numbers in any order

$50 \times 4 = 4 \times 50$

$60 + 30 = 30 + 60$

The **Commutative Property** makes arithmetic easier.

See also: **Associative Property, Distributive Property**

compare /kuhm PAIR/ (v.)

1. to figure out which number in a pair is greater

Use the symbols <, >, or = to **compare** the numbers.

$6 > 4$ — 6 is greater than 4.

$-8 < -5$ — −8 is less than −5.

$\frac{1}{4} = \frac{2}{8}$ — $\frac{1}{4}$ is equal to $\frac{2}{8}$.

2. to note the difference or sameness of two things

Compare the weights of a mouse and a cat. Which weighs more?

See also: **greater than, less than, order, symbol**

compass /KUM puhs/ (n.)

1. a tool for drawing circles
   and arcs

Use a **compass** to draw a circle or arc.

2. a tool for finding a geographic
   direction

See also: **arc, circle, compass points, construct**

The needle on a **compass** always points to the north.

compass points
/KUM puhs points/ (n.)

directions on a compass,
such as north, south, east,
and west

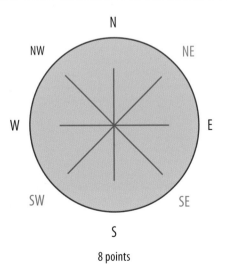

8 points

See also: **compass**

The **compass points** N and S are opposite
of each other.

complementary angle
/kom pluh MEN tur ee ANG guhl/ (n.)

one of two angles whose sum is 90°

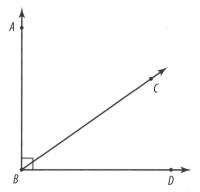

See also: **right angle, straight angle, supplementary angle**

Angle *ABC* and angle *CBD* are
complementary angles.
m∠*ABC* + m∠*CBD* = 90°

---

concave /kon KAYV/ (adj.)

1. a polygon with an angle greater than 180°

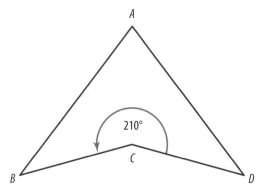

Angle *BCD* measures 210°. Quadrilateral *ABCD* is a **concave** polygon.

2. curving inward

See also: **convex, polygon, reflex angle, vertex**

The inside of a spoon is **concave**. It can hold water.

concentric /kuhn SEN trik/ (adj.)

with the same center

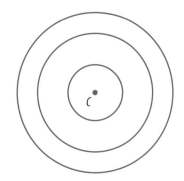

See also: **center, circle, intersect**

The three circles are **concentric**. They share the center point C. The circles do not intersect.

cone /kohn/ (n.)

a solid with a circle base and one vertex

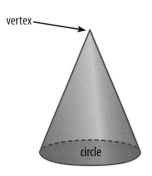

See also: **base, circle, solid, vertex**

The base of a **cone** is a circle.

congruent (≅) /kuhn GROO uhnt/ (adj.)

having the same size and shape

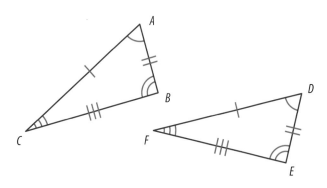

$$\angle A \cong \angle D \qquad \angle C \cong \angle F$$
$$\overline{AB} \cong \overline{DE} \qquad \overline{BC} \cong \overline{EF}$$

See also: **angle, side, triangle**

Triangle ABC is **congruent** to triangle DEF.

consecutive
/kuhn SEK yuh tiv/ (adj.)

one after the other in order

12, 13, 14, 15, 16, 17

These are **consecutive** whole numbers.

7, 9, 11, 13, 15, 17

These are **consecutive** odd numbers.

See also: **odd, order, whole number**

---

construct /kuhn STRUKT/ (v.)

to draw a geometric figure
with a compass and ruler

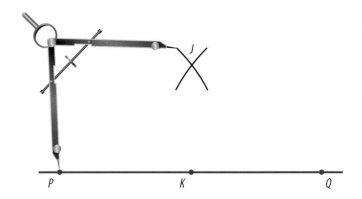

See also: **compass, line segment,
perpendicular, ruler**

You can **construct** a line segment perpendicular to $\overline{PQ}$.
The line segment $\overline{JK}$ will be perpendicular to $\overline{PQ}$.

---

convert /kuhn VURT/ (v.)

to change from one unit
to another

You can **convert** meters to centimeters:

$$2\ m = 2 \times 100\ cm$$
$$= 200\ cm$$

There are 200 centimeters in two meters.

See also: **centimeter, meter**

convex /kon VEKS/ (adj.)

1. a polygon with each angle less than 180°

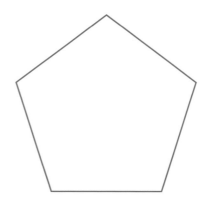

This a **convex** polygon.

2. curved outward

The outside of a measuring spoon is **convex**.
It cannot hold water.

See also: **concave, polygon, vertex**

---

coordinate grid
/koh ORD nit grid/ (n.)

a graph for finding points

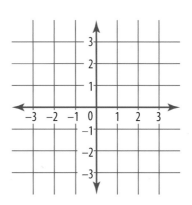

A **coordinate grid** has an *x*-axis and *y*-axis. The two axes meet at the origin, or (0, 0). A **coordinate grid** has four quadrants.

See also: **axis, ordered pair, origin, quadrant**

coordinates /koh ORD nits/ (n.)

numbers or letters that describe
a position

See also: **coordinate grid, ordered pair,
x-coordinate, y-coordinate**

The **coordinates** of A are (1, 3). The **coordinates** of B
are (3, 2).

copy /KOP ee/ (v.)

to make something new that is
the same as something else

Copy a line segment with
a compass and ruler.

See also: **compass, construct, line segment**

copy /KOP ee/ (n.)

something that is exactly
the same as something else

See also: **reflection, rotation,
transformation, translation**

Triangle B is a **copy** of triangle A.

corner /KOR nur/ (n.)

the point where sides
or edges meet

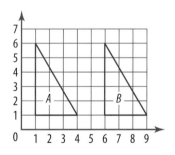

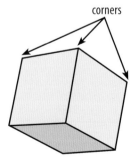

See also: **cube, edge, side, vertex**

A **corner** is another name for a vertex. A cube has
eight **corners**. A square has four **corners**.

correct /kuh REKT/ (adj.)
with no mistakes or errors

What is 5 + 6?
The **correct** answer is 11.

See also: **accurate, answer, check, solution**

---

correct /kuh REKT/ (v.)
to fix a mistake or error

10 + 25 = ~~45~~

I have to **correct** this mistake in my homework.
The answer is 35, not 45.

See also: **answer, check, solution**

---

cost /kawst/ (n.)
the amount paid for something

See also: **pay, per, pound, price**

The **cost** of the carrots is $1.50 per pound.

---

count /kownt/ (v.)
1. to say numbers in order

The students **count** from 1 to 5.
1, 2, 3, 4, 5
They then **count** by 5s from 5 to 25.
5, 10, 15, 20, 25

2. to add up; to find the total number in a group of something

I **count** the sides of a triangle. I **count** three sides. The triangle has three sides.

See also: **ascending, order, side, triangle**

counter /KOWN tur/ (n.)

an object that helps you count
or do arithmetic

See also: **arithmetic, count**

There are eight red **counters**. There are four
green **counters**. There are twelve **counters** in all.

counterclockwise
/kown tur KLOK wīz/ (adv.)

in the opposite direction that
the hands of a clock turn

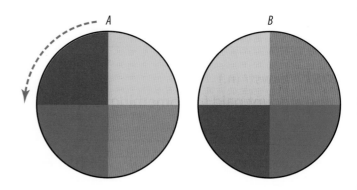

See also: **clockwise, minute hand,
opposite, rotate**

Rotate circle *A* **counterclockwise** a quarter turn.
The result is circle *B*.

cover /KUV ur/ (v.)

to put over, or on top of,
something else

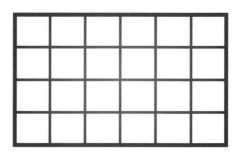

See also: **area, rectangle, square unit**

Find how many square units completely **cover**
the rectangle. This is the area of the rectangle.

cross /kraws/ (v.)

to intersect, to share a point

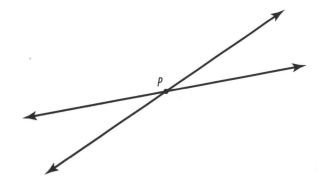

See also: **intersect, line, point**

The two lines **cross** at point *P*.

---

cross-section
/kraws SEK shuhn/ (n.)

the two-dimensional shape that
is made by cutting through
a solid figure

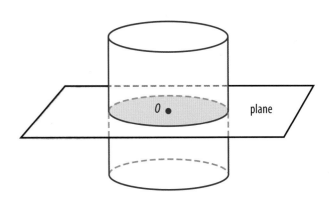

See also: **intersect, plane, solid,
two-dimensional**

The plane intersects the cylinder. The **cross-section** of
the cylinder is a circle. The center of the **cross-section**
is point *O*.

---

cube /kyoob/ (n.)

a solid with six square faces

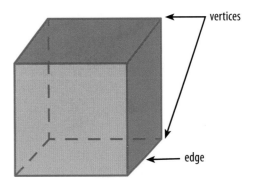

See also: **face, solid, square,
three-dimensional**

A **cube** has 12 edges and 8 vertices.

cube /kyoob/ (n.)

a base with an exponent of 3; the third power

$$2^3 = 2 \times 2 \times 2$$
$$= 8$$

The **cube** of 2 is 8.

See also: **base, exponent, power, square**

cube root /kyoob root/ (n.)

a number that is multiplied three times to equal another number

The **cube root** of 27 is 3.
$27 = 3 \times 3 \times 3 = 3^3$

The **cube root** of 1,000 is 10.
$1,000 = 10 \times 10 \times 10 = 10^3$

See also: **base, cube, exponent, power**

cubic unit /KYOO bik YOO nit/ (n.)

a quantity for measuring volume or capacity

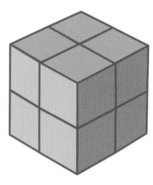

The volume of the cube is 8 **cubic units**.
Cubic centimeters (cm^3), cubic inches (in^3), and cubic feet (ft^3) are **cubic units**.

See also: **capacity, cube, measure, volume**

cup /kup/ (n.)

a measure of capacity in the customary system

8 ounces = 1 **cup**
2 **cups** = 1 pint
4 **cups** = 1 quart

See also: **capacity, customary system, ounce, pint, quart**

curve /kurv/ (n.)

1. a line that bends or that is not straight

This is a **curve** because it bends.

2. a line on a graph that shows data

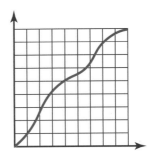

See also: **data, graph, line**

The **curve** makes the data easier to understand.

---

customary system
/KUS tuh mair ee SIS tuhm/ (n.)

the most common measurement system in the United States

See also: **measurement, metric system**

Customary System Units		
Weight	**Length**	**Capacity**
ounces	inches	cups
pounds	feet	pints
tons	yards	quarts
	miles	gallons

---

cylinder /SIL uhn dur/ (n.)

a solid with two circular faces and one curved surface

The two faces of a **cylinder** are circles. A curved surface connects the two circles.

See also: **circle, face, solid, surface**

Dd

D /dee/ (n.)

the Roman numeral representing 500

D = 500	
DL = 550	
DC = 600	

See also: **C, L, numeral, Roman numerals**

D /dee/ (abbr.)

dimension; having length, width, depth, or all three

A circle is a two-dimensional, or 2-**D**, figure.
A cylinder is a three-dimensional, or 3-**D**, figure.

See also: **circle, cylinder, three-dimensional, two-dimensional**

---

data /DAY tuh/ (n.)

information that is numbers, words, or both

Favorite Colors	
Color	**Number of Students**
Blue	25
Red	22
Yellow	6
Green	11
Purple	24

Lena collected **data**. She organized her **data** in this table.

See also: **chart, database, graph, table**

---

database /DAY tuh bays/ (n.)

a set of data stored and organized on a computer

Use a **database** to sort numbers.
You can sort in ascending or descending order.

See also: **ascending, data, descending**

date /dayt/ (n.)
   the day, month, and year

See also: **calendar, month**

July 4, 1776
11/14/1996
3-26-85

You can write a **date** in different ways.

---

day /day/ (n.)
   a period of twenty-four hours

24 hours = 1 **day**
7 **days** = 1 week

See also: **hour, month**

---

decagon /DEK uh gon/ (n.)
   a polygon with ten sides

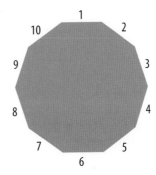

See also: **polygon, regular polygon, side**

This regular **decagon** has ten sides. The sides all have the same length.

---

decahedron /dek uh HEE druhn/ (n.)
   a solid with ten faces

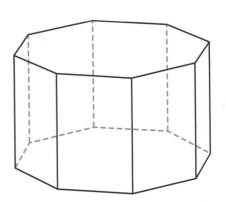

See also: **base, face, prism, solid**

A **decahedron** has ten faces. *Deca-* means "ten."

35

deciliter (dl) /DES uh lee tur/ (n.)

a unit of volume equal
to one tenth of a liter

$1\ \textbf{deciliter} = \frac{1}{10}\ \text{liter}$
$10\ \textbf{deciliters} = 1\ \text{liter}$
$1\ \textbf{deciliter} = 100\ \text{milliliters}$
Deci- means "one tenth."

See also: **capacity, liter, metric system, milliliter**

decimal /DES uh muhl/ (n.)

a number with a decimal point
in the base-ten system

This is a **decimal**:

22.47

It is equal to $20 + 2 + 0.4 + 0.07$. A **decimal** has digits to the right of the decimal point.

See also: **decimal place, decimal point, expanded form, whole number**

decimal place
/DES uh muhl plays/ (n.)

a digit's position to the right
of the decimal point

What is the **decimal place** of each digit in this number?

3.745

thousandths' place
hundredths' place
tenths' place

See also: **decimal point, digit, place value, whole number**

A B C **D** E F G H I J K L M N O P Q R S T U V W X Y Z   **decrease**

decimal point
/DES uh muhl point/ (n.)

1. a dot between the one tenths' place and the ones' place in a decimal

4͓2.6

Read this number as "forty-two point six" or "forty-two and six tenths."
Read the **decimal point** as "and" or "point."

2. a dot between the dollars and cents in money amounts

$10͓.50

Cents are to the right of the **decimal point**.

See also: **decimal, decimal place, digit**

decimeter (dm)
/DES uh mee tur/ (n.)

a unit of length equal to one tenth of a meter

1 **decimeter** = $\frac{1}{10}$ meter
10 **decimeters** = 1 meter
1 **decimeter** = 10 centimeters
Deci- means "one tenth."

See also: **centimeter, length, meter, metric system**

decrease /di KREES/ (v.)
to make less or smaller

The cars are traveling 65 miles per hour (mph).
They **decrease** their speed by 15 mph.
They are now traveling 50 mph.

See also: **increase, speed**

decrease /di KREES/ (n.)
the amount of change that makes something less or smaller

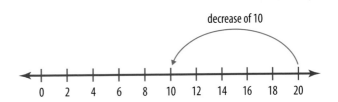
decrease of 10

The temperature was 20 degrees at 7:00 p.m. It was 10 degrees at 3:00 a.m. There was a **decrease** in temperature of 10 degrees.

See also: **increase**

37

degree /di GREE/ (n.)

1. a unit used for measuring temperature

> The temperature is 25 **degrees** Fahrenheit. It is cold.
> The temperature is 85 **degrees** Fahrenheit. It is hot.

2. a unit used for measuring angles

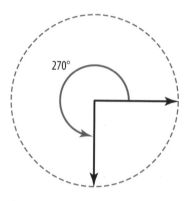

There are 360 **degrees** in a circle.

See also: **angle, circle, measure, temperature**

denominator
/di NOM uh nay tur/ (n.)

the bottom number of a fraction

numerator

$$\frac{2}{3}$$

denominator

The **denominator** tells the total number of equal parts. The **denominator** is the divisor. It cannot equal 0.

See also: **dividend, divisor, fraction, numerator**

dependent variable
/di PEN duhnt VAIR ee uh buhl/ (n.)

a variable with a value based on the value of another variable

$$d = 40 \times t$$

The **dependent variable** is d. Here d is distance and t is time. The distance you drive depends on the time you spend driving.

See also: **independent variable, value, variable, x**

depth /depth/ (n.)

1. a measure of how deep something is

2. one of the three dimensions of a three-dimensional object

The **depth** of the water is 33 feet.

depth

See also: **dimension, height, three-dimensional, width**

The **depth** of the box is from front to back. The box also has height and width.

---

descending /di SEND ing/ (adj.)

becoming lesser in size or value

These numbers are in **descending** order:

425, 305, 222, 100, 55, 12, 3

See also: **ascending, order**

---

diagonal /dī AG uh nuhl/ (n.)

a line joining opposite corners in a shape

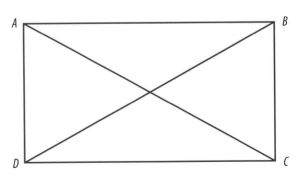

$\overline{AC}$ is a **diagonal** of rectangle *ABCD*.
$\overline{BD}$ is also a **diagonal**.
$\overline{AC}$ and $\overline{BD}$ are the same length.

See also: **corner, polygon, vertex**

39

diagram /DĪ uh gram/ (n.)

a picture that makes something more clear or easier to understand

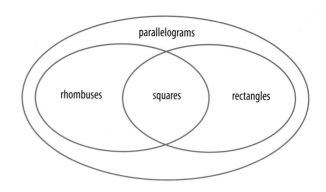

This **diagram** shows the different kinds of quadrilaterals. It uses circles or ovals to show how the different figures are related.

See also: **parallelogram, quadrilateral, rectangle, Venn diagram**

diameter /dī AM uh tur/ (n.)

a line segment in a circle or sphere; it has endpoints on the circle and goes through the center

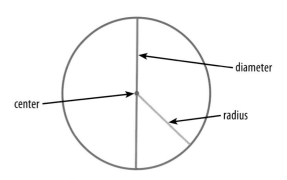

See also: **circle, radius, sphere**

A circle's **diameter** is twice the radius.

diamond /dī muhnd/ (n.)

a shape with four equal sides and no right angles

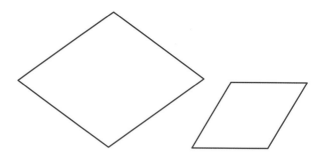

See also: **equal, rhombus, right angle, square**

A **diamond** is a kind of rhombus. These two figures are **diamonds**.

difference   /DIF ur uhns/ (n.)

the amount left when one number is subtracted from another

$$20 - 9 = 11$$

The **difference** between 20 and 9 is 11.

See also: **addition, subtraction, sum**

---

different   /DIF ur uhnt/ (adj.)

not the same; unlike

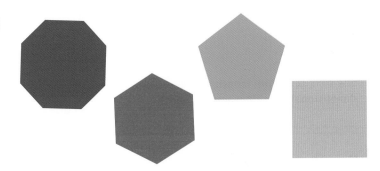

See also: **compare**

Each shape is a **different** color. Each has a **different** number of sides.

---

digit   /DIJ it/ (n.)

any numeral from 0 to 9

1,457

This number has four **digits**. The ones' **digit** is 7.

See also: **base, number, numeral**

---

digital   /DIJ uh tuhl/ (adj.)

showing the time with numbers

See also: **clock, hour hand, minute hand**

A **digital** clock does not have hands. Read a **digital** clock by reading the numbers.

41

dimension /duh MEN shuhn/ (n.)

a measure of width, length, or height

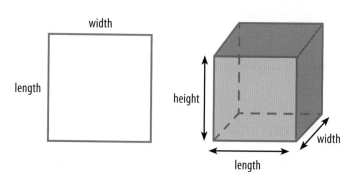

See also: **D, height, length, three-dimensional, two-dimensional, width**

A square has two **dimensions**: length and width.
A cube has three **dimensions**: length, width, and height.

direct proportion
/duh REKT pruh POR shuhn/ (n.)

a relation where two or more quantities change in the same way

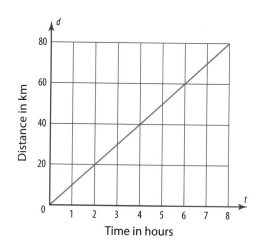

See also: **decrease, increase, inverse proportion**

The distance, d, increases as the time, t, increases.
These two quantities are in **direct proportion**.

discount /DIS kownt/ (n.)

an amount taken away from the usual price of something

See also: **cost, percent, price**

The price of a shirt is $20. There is a **discount** of ten percent, or $2. The new price, or sale price, is $18.

distance /DIS tuhns/ (n.)

the amount of space between things or places

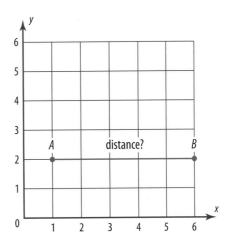

1. The **distance** from point A to point B is 5 units.
2. You can also find the **distance** an object travels. Use the formula $d = rt$, where d = **distance**, r = rate, and t = time.

See also: **coordinate grid, point, rate, time**

Distributive Property
/dis TRIB yuh tiv PROP ur tee/ (n.)

the rule that says a number times a sum is equal to the sum of the products

There is a **Distributive Property** of Multiplication over Addition.
$$8 \times 25 = 8 \times (20 + 5)$$
$$= (8 \times 20) + (8 \times 5)$$

There is also a **Distributive Property** of Multiplication over Subtraction.
$$12 \times (10 - 2) = (12 \times 10) - (12 \times 2)$$

See also: **multiplication, product, sum**

divide (÷ , / , ⌐) /duh VĪD/ (v.)

1. to separate into equal parts or groups

Three students **divide** twelve cookies evenly among them.
Each student gets four cookies.

2. to find how many times one number goes into another number

$$30 / 3 = 10$$
$$30 \div 3 = 10$$

Divide thirty by three to get ten. Ten **divides** thirty three times. Three goes into thirty ten times.

See also: **division, part, whole**

43

dividend /DIV uh dend/ (n.)

the quantity divided
in a division problem

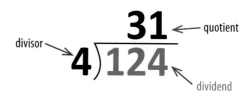

In this division problem, 124 is the **dividend**.

See also: **division, divisor, quotient**

---

divisible /duh VIZ uh buhl/ (adj.)

can be divided by another
number without a remainder

$45 \div 9 = 5$

Forty-five is **divisible** by nine. There is no remainder. Nine goes into forty-five exactly five times.

See also: **division, remainder**

---

division (÷, /, ⌐)
/duh VIZH uhn/ (n.)

the act of finding how many
times one number goes into
another number

You can use an array to show **division**.
Twelve divided by three is four.
$12 \div 3 = 4$

See also: **array, divide, multiplication, remainder**

---

divisor /duh VĪ zur/ (n.)

a number that is divided into
another number or quantity

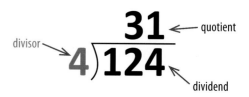

In this division problem, 4 is the **divisor**.

See also: **dividend, division, quotient**

dodecagon   /doh DEK uh gon/ (n.)

a polygon with twelve sides

See also: **polygon, regular polygon, side**

This regular **dodecagon** has twelve sides.
The sides all have the same length.

dodecahedron

/doh dek uh HEE druhn/ (n.)

a solid with twelve faces

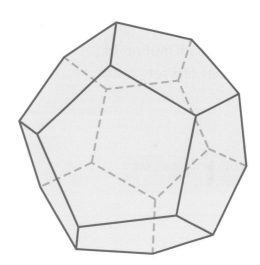

See also: **face, net, solid**

Each face of this **dodecahedron** is a pentagon.

double   /DUB uhl/ (v.)

to make twice as much or as many

Double 15 to get 30. To **double** a number:
Multiply the number by 2. $15 \times 2 = 30$
Add the number to itself. $15 + 15 = 30$

See also: **add, multiply**

edge /ej/ (n.)

the place where two surfaces
of a solid meet

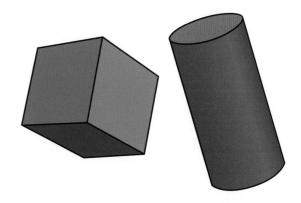

See also: **cube, cylinder, face, solid**

An **edge** can be straight or curved. A cube has twelve
edges. A cylinder has two curved **edges**.

element /EL uh muhnt/ (n.)

one member of a set

Five is an **element** of the set of whole numbers.
An **element** of the set of rational numbers is $\frac{5}{10}$.

See also: **rational number, set, whole number**

ellipse /i LIPS/ (n.)

a plane shape that is a
cross-section of a cone

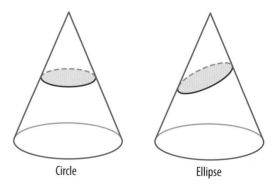

Circle   Ellipse

See also: **cone, cross-section, line of symmetry, oval**

An **ellipse** looks like an oval or a stretched circle.
An **ellipse** has two lines of symmetry. A circle is
a special kind of **ellipse**.

empty /EMP tee/ (adj.)
containing nothing

A container that has nothing in it is **empty**.

A set with no elements is an **empty** set. These symbols show that a set is **empty**: { } and Ø

See also: **capacity, element, set**

---

endpoint /END point/ (n.)
a point at the end of a line segment, ray, or arc

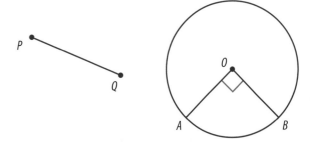

See also: **arc, line segment, point, ray**

The name of the line segment is $\overline{PQ}$. P is an **endpoint** of line segment $\overline{PQ}$. The name of the arc is $\overset{\frown}{AB}$.

---

enlarge /en LARJ/ (v.)
to make bigger

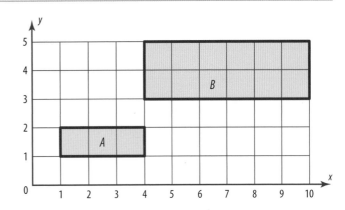

See also: **dimension, reduce, scale, times**

Multiply each side length by two to **enlarge** rectangle A. Rectangle A was **enlarged**. Rectangle B is the result.

---

equal /EE kwuhl/ (adj.)
being the same in size, degree, or amount

Four cups is **equal** to one quart.

One hundred centimeters is **equal** to one meter.

See also: **centimeter, cup, equivalent, quart**

equals sign (=) /EE kwuhlz sīn/ (n.)

the symbol that shows two things
have the same value

An **equals sign** shows the two sides
of an equation are the same.

3 × 5 = 15

Read this as, "Three times five equals fifteen."

See also: **equal, equation, equivalent**

---

equation /i KWAY zhuhn/ (n.)

two expressions with an
equals sign between them

3 × 10 = 30
100 + x = 175
2b + 4 = b + 7

Two sides of an **equation** have the same value.
Equations can have numbers, variables, or both.

See also: **equal, equals sign, inequality,
variable**

---

equilateral triangle
/ee kwuh LAT ur uhl TRĪ ang guhl/ (n.)

a triangle with all sides
the same length

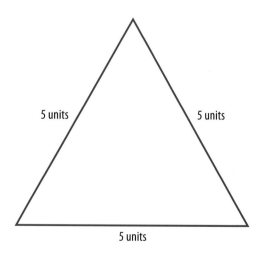

The three angles in an **equilateral triangle** are
also equal. Each angle measures 60 degrees.

See also: **angle, degree, equal, triangle**

equivalent   /i KWIV uh luhnt/ (adj.)

having the same value, worth, or meaning

Two nickels are **equivalent** to one dime. Both have the value of ten cents.

See also: **equal, equivalent fraction, value**

---

equivalent fraction
/i KWIV uh luhnt FRAK shuhn/ (n.)

a fraction that has the same value as another fraction

$$\frac{4}{6} = \frac{2}{3} \qquad \frac{3}{4} = \frac{75}{100}$$

To find an **equivalent fraction**, multiply or divide.

For example, $\frac{3}{4}$ is also equivalent to $\frac{6}{8}$.

$$\frac{3}{4} = \frac{2\times3}{2\times4} = \frac{6}{8}$$

See also: **equivalent, simplify, value**

Eratosthenes' sieve
/air uh TAHS thuh neez siv/ (n.)

a way to find prime numbers less than a certain number

Rules for an **Eratosthenes' sieve**:
1. To find prime numbers up to 50, make a chart of the numbers 1 to 50.
2. Circle 2. It is the least prime number. Cross out all multiples of 2.
3. Circle 3, or the next prime number. Cross out all multiples of 3.
4. Circle the next number not crossed out, or 5. Cross out all multiples of 5.
5. Stop when all numbers are crossed out or circled. The circled numbers are prime numbers.

See also: **factor, multiple, prime number**

estimate /ES tuh mayt/ (v.)
to find an answer close to
an exact answer

To **estimate** 48 × 36, round up to 50 × 40.

See also: **approximate, exact**

estimate /ES tuh mit/ (n.)
an answer close to
an exact answer

I make an **estimate** that about 25% of
students at our school have pets. I asked
100 random students about pets. Exactly
24 said they had pets. My **estimate**
was close.

See also: **answer, estimation, exact,
random**

estimation
/es tuh MAY shuhn/ (n.)
the process of finding an answer
close to an exact answer

Multiply 1.9 and 4.2. You can use **estimation**
to check your answer.

Calculation: **Estimation:**
1.9 × 4.2 = 7.98   2 × 4 = 8

Your calculation is probably correct.
Your **estimation** shows that 2 times 4 is 8.

See also: **answer, approximate,
calculation, exact**

even /EE vuhn/ (adj.)
divisible by two with
no remainder

6

See also: **divisible, odd, remainder,
whole number**

Six is an **even** number. The opposite of **even**
is *odd*.

event /i VENT/ (n.)

something that can happen in a probability experiment

See also: **outcome, probability, random, trial**

Mr. Salazar flips two coins twenty times.
This experiment has three possible **events**.
The possible **events** are:
1. two heads,
2. two tails, or
3. one head and one tail.

---

exact /eg ZAKT/ (adj.)

very precise or accurate;
not approximate in any way

The **exact** sum of 6.2 + 7.6 is 13.8.

See also: **accurate, approximate, estimate, length**

---

exchange /eks CHAYNJ/ (v.)

to change for something else
that is equivalent

I want coins for my one-dollar bill. I can **exchange**
the bill for these different coins.
• 3 quarters, 2 dimes, 1 nickel
• 2 quarters, 4 dimes, 2 nickels
• 8 dimes, 20 pennies

See also: **change, coin, value**

expanded form
/ek SPAND uhd form/ (n.)

a number shown as the sum
of the values of the digits

$$(2 \times 10,000) + (3 \times 1,000) + (4 \times 100) + (1 \times 10) + (5 \times 1)$$

This is the number 23,415 in **expanded form**.

See also: **digit, standard form, sum, value**

exponent /ek SPOH nuhnt/ (n.)

the number a base is raised to;
the power

$$8^3 = 512$$

The number 8 is the base raised to the power of 3.
The 3 is the **exponent**.
$8 \times 8 \times 8 = 512$

See also: **base, factor, power**

expression /ek SPRESH uhn/ (n.)

a math phrase with numbers,
variables, and/or symbols

$2 + 5$

$6x$

$10a + 20b + 4$

Each example above is an **expression**.

See also: **equation, inequality, symbol,
variable**

Ff

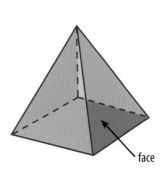

face

face

face /FAYS/ (n.)
a flat surface on a solid

A square pyramid has one square **face**. It has four triangular **faces**. A cylinder has two circular **faces**.

See also: **base, edge, solid, vertex**

---

factor /FAK tur/ (n.)
a number that is multiplied

$$5 \times 20 = 100$$

Multiply 5 and 20 to get 100.
5 is a **factor** of 100. 20 is another **factor**.
The product is 100.

See also: **divisor, prime factor, prime number, product**

Fahrenheit /FAIR uhn hīt/ (n.)
a scale for measuring temperature

See also: **Celsius, temperature**

Water freezes at 32° in the **Fahrenheit** scale.
Water boils at 212° in the **Fahrenheit** scale.

few /fyoo/ (adj.)
a small number of something

She has three or four coins in her pocket.
She has **few** coins.

See also: **coin**

figure /FiG yur/ (n.)

1. a plane shape made by one
or more lines or curves

Each **figure** in the drawing has four sides. The two **figures** are congruent.

2. a diagram or picture that
makes something more clear

Figure 2 shows the company made more money each year.

3. a number, a digit, or an amount
of money

Company Profits (in millions)				
	Quarter 1	Quarter 2	Quarter 3	Quarter 4
2007	$2.3	$2.3	$2.6	$2.9
2008	$3.0	$3.1	$3.5	$3.6
2009	$3.8	$4.1	$3.9	$4.0
2010	$4.1	$4.1	$4.3	$4.4

The **figures** in the table show the company makes more money each year.

See also: **congruent, diagram, digit**

figure /FIG yur/ (v.)
to calculate or work out
a solution

To **figure** a rectangle's area, find the product
of its length and width.

See also: **calculate, solution, solve**

---

find /fīnd/ (v.)
to calculate or figure out

Add $1.50 and $4.00 to **find** the solution.

See also: **calculate, cost, solution, solve**

---

flat /flat/ (n.)
a base-ten block that represents
one hundred

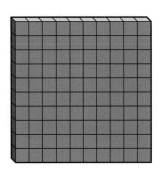

1 **flat** = 10 longs
1 **flat** = 100 cubes
These are two **flats**. They represent 200
in the base-ten system.

See also: **cube, long**

---

flat /flat/ (adj.)
smooth and even

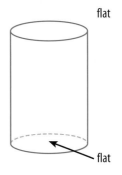
flat

flat

A cylinder has two **flat** surfaces. They are circles.
The cylinder has one curved surface.

See also: **curve, cylinder, round**

55

flip /flip/ (v.)

1. to turn over

A ruler has inches on one side and centimeters on the other side. **Flip** the ruler over to use the centimeter side.

2. to toss a coin

Flip a coin 50 times. How many times does the coin land heads up?

3. to reflect

Flip figure *A* over the dashed line. The result is figure *B*. The two figures are congruent.

See also: **coin, congruent, reflect, ruler**

flip /flip/ (n.)

another name for a reflection

See also: **congruent, reflection, rotation, translation**

Figure *B* is a **flip** of figure *A*. The two figures are congruent.

flow chart /floh chart/ (n.)

a picture with shapes and arrows that shows actions

The Order of Operations

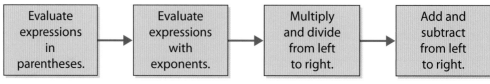

See also: **diagram, order of operations**

This **flow chart** shows the order of operations.

fold /fohld/ (v.)

to bend something so that one
part covers another part

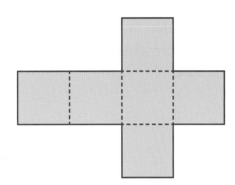

This net makes a cube. **Fold** on the dashed lines.
Cut on the solid lines.

See also: **cover, cube, net**

---

foot /fut/ (n.)

a unit of length in
the customary system

A ruler measures one **foot**.
1 **foot** = 12 inches
3 **feet** = 1 yard
5,280 **feet** = 1 mile

See also: **customary system, foot, mile, yard**

---

formula /FOR myuh luh/ (n.)

a rule for finding a value

The **formula** for the area of a triangle
is $A = \frac{1}{2}(b \times h)$.
In the **formula**, b = measure of base
and h = measure of height.

See also: **area, rule, triangle, value**

---

forward /FOR wurd/ (adv.)

in the usual way or direction

10, 11, 12, 13, 14, 15, 16, 17, 18,
19, 20

You can count **forward** from 10 to 20. You begin
at 10 and end at 20.

See also: **backward, count**

fraction /FRAK shuhn/ (n.)

1. a number that represents a part of a whole

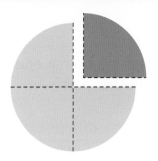

This circle has four equal parts.

$\frac{3}{4}$ is the **fraction** that is light blue.

$\frac{1}{4}$ is the **fraction** that is darker blue.

2. a number that shows one number divided by another

The **fraction** $\frac{1}{2}$ also means 0.5.

Divide 1 by 2 to get 0.5, or $\frac{1}{2}$.

Both $\frac{1}{2}$ and 0.5 are quotients.

3. a ratio comparing two quantities

See also: **decimal number, improper fraction, quotient, ratio**

Two out of five apples are bad.

The **fraction** comparing bad apples to good apples is $\frac{2}{5}$.

frequency /FREE kwuhn see/ (n.)

how often a value happens in a data set

Grade	Number of Students, or Frequency
90–100	IIII
80–89	HHT HHT II
70–79	HHT III
60–69	II

See also: **data, set, tally, value**

The tally marks show the **frequency** of grades.

full /ful/ (adj.)

containing as much as possible

See also: **empty**

This measuring cup is **full** to the top line.

Gg

gallon (gal.) /GAL uhn/ (n.)
a unit of volume in the customary system

1 **gallon** = 8 pints
1 **gallon** = 4 quarts
1 **gallon** = 16 cups

See also: **capacity, cup, customary system, pint, quart**

geoboard /JEE o bord/ (n.)
a flat surface with pegs set in a rectangular grid; used to make shapes

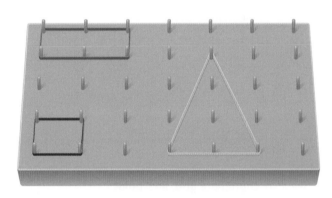

See also: **polygon, rectangle, square, triangle**

Use a **geoboard** to make polygons. There is a triangle, a square, and a rectangle on this **geoboard**.

geometry /jee OM uh tree/ (n.)
the study of points, lines, planes, shapes, and solids

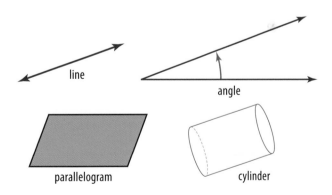

line

angle

parallelogram

cylinder

See also: **line, plane, point, rectangle, solid**

You learn about lines, shapes, and solids in **geometry** class.

59

gram (g) /gram/ (n.)

a unit of mass in
the metric system

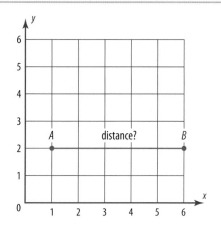

A raisin and a paper clip each weighs about one **gram**.
1 **gram** = 1,000 milligrams
1 **gram** = 0.001 kilogram
1,000 **grams** = 1 kilogram

See also: **kilogram, mass, metric system, weight**

---

graph /graf/ (v.)

to draw a point or a line
on a coordinate grid

See also: **coordinate grid, line, point**

First, plot point A. Then, plot point B. Last, **graph** the line segment.

---

graph /graf/ (n.)

plotted points on
a coordinate grid

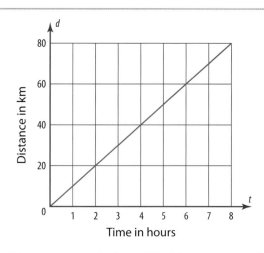

See also: **coordinate grid, line, point**

This is a **graph** of a line. All of the points on the **graph** follow the rule $d = 10t$, where d is distance traveled and t = time, in hours.

greater than (>)

/GRAY tur THan/ (adj.)

more than

Use the symbol > to compare two numbers.

100 is **greater than** 99　　　100 > 99

0.75 is **greater than** 0.50.　0.75 > 0.50

See also: **equal, less than, symbol**

---

greatest common factor (GCF)

/GRAYT est KOM uhn FAK tur/ (n.)

the largest whole number that is a factor of all the numbers in a set

To find the **greatest common factor** of two numbers:

1. List the factors of each number.

   Factors of 16: 1, 2, 4, 8, 16

   Factors of 24: 1, 2, 3, 4, 6, 8, 12, 24

2. Find the common factors and choose the greatest one: 1, 2, 4, <u>8</u>.

See also: **factor, prime factor, product, whole number**

---

group /groop/ (v.)

to put together by kind or category; to sort

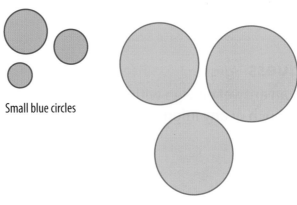

Small blue circles

Large red circles

You can **group** figures in many ways. They can be **grouped** by shape. They can also be **grouped** by size or color.

group /groop/ (n.)

a number of things that are close together or have something in common

There are many kinds of buttons in this **group**.

---

guess /ges/ (v.)

to choose an answer without much information

I don't know this town well. I **guess** it is about two miles to the nearest store.

See also: **answer, estimate, exact**

---

guess /ges/ (n.)

an answer chosen without much or any information

Mr. Bridges is thinking of a number between 1 and 100. Maria's **guess** is that the number is 45.

See also: **answer, estimate, exact**

half /haf/ (n.)
one of two equal parts

See also: **equal, part, whole**

Half of the triangle is orange. **Half** of the 16 squares are blue.

heavy /HEV ee/ (adj.)
of a large amount of weight

The car is very **heavy**. It weighs two tons.
A leaf is not **heavy**. It weighs only one gram.

See also: **gram, mass, ton, weight**

height /hīt/ (n.)
1. the shortest line segment that can join a corner and opposite side

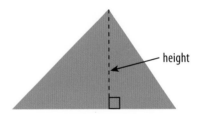

The dashed line is the **height** of the triangle.
The **height** and the base form a right angle.

2. one of the three dimensions of a three-dimensional object

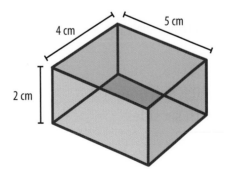

See also: **altitude, length, three-dimensional, width**

The **height** of the rectangular prism is 2 cm.
Its width is 4 cm. Its length is 5 cm.

hemisphere /HEM uh sfeer/ (n.)
half of a sphere

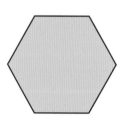

circle

See also: **circle, face, half, sphere**

A **hemisphere** has one flat face. This face is a circle.

heptagon /HEP tuh gon/ (n.)
a polygon with seven sides

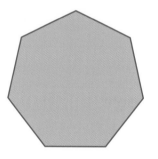

This regular **heptagon** has seven sides.
The sides all have the same length.

See also: **polygon, regular polygon, side**

This irregular **heptagon** also has seven sides.
The sides are not all the same length.

hexagon /HEK suh gon/ (n.)
a polygon with six sides

See also: **polygon, regular polygon, side**

A **hexagon** has six sides and six angles.

hexagonal prism
/hek **SAG** uh nuhl **PRIZ** uhm/ (n.)

a prism made from two bases that are hexagons

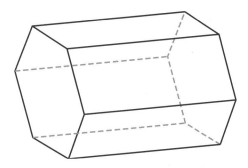

See also: **base, face, hexagon, prism**

The two bases of a **hexagonal prism** are hexagons. The other six faces are rectangles.

hexahedron
/hek suh **HEE** druhn/ (n.)

a solid with six faces

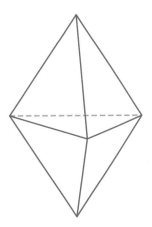

This is a picture of a **hexahedron**. It has six faces that are triangles.

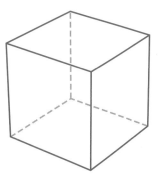

See also: **cube, face, solid**

This is a picture of a cube, another kind of **hexahedron**. It has six faces that are squares.

65

horizontal /hor uh ZON tl/ (adj.)

parallel to the horizon;
at a right angle to vertical

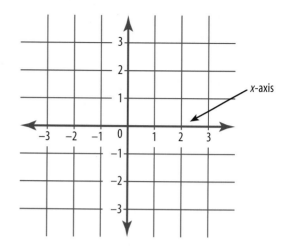

x-axis

See also: **line, perpendicular, right angle, vertical**

The x-axis is a **horizontal** line.

hour /owr/ (n.)

a unit of time equal
to sixty minutes

1 **hour** = 60 minutes
24 **hours** = 1 day

See also: **day, minute, time**

hour hand /owr hand/ (n.)

the hand on the clock that shows
the hour

hour hand

See also: **clock, hour, minute hand**

The **hour hand** is shorter than the minute hand.

Ii

I /ī/ (n.)
the Roman numeral that
represents the number 1

II = 2; 1 + 1 = 2
III = 3; 1 + 1 + 1 = 3
VII = 7; V = 5, and 5 + 1 + 1 = 7
XIII = 13; X = 10, and 10 + 1 + 1 + 1 = 13

See also: **Roman numerals**

icosahedron
/ī cohs uh HEE druhn/ (n.)
a solid with 20 faces

See also: **face, polyhedron, solid, triangle**

Each face of a regular **icosahedron** is a triangle.

identical /ī DEN tuh kuhl/ (adj.)
exactly the same shape and size

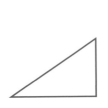

The shapes in the picture are all **identical**.

impossible
/im POS uh buhl/ (adj.)
not able to happen

The bag only has
blue counters.
It is **impossible** to
pull a green counter
from the bag.

See also: **probability**

improper fraction
/im PROP ur FRAK shuhn/ (n.)

a fraction with a value greater than 1; a fraction in which the numerator is greater than the denominator

See also: **denominator, fraction, numerator**

An example of an **improper fraction** is $\frac{9}{4}$.

Its value can also be shown as $2\frac{1}{4}$.

$$\frac{4}{4} \qquad \frac{4}{4} \qquad \frac{1}{4}$$

$$\frac{8}{4}$$

$$\frac{9}{4}$$

---

inch /inch/ (n.)

a small unit of length in the customary system

one inch

Inches (in.)

See also: **foot, measure, unit**

An adult's big toe is about one **inch** long.
12 **inches** = 1 foot

---

increase /in KREES/ (v.)

to make greater or larger

The cars are traveling 50 miles per hour (mph).
They **increase** their speed by 15 mph.
They are now traveling 65 mph.

See also: **add, decrease**

---

increase /IN krees/ (n.)

the amount of change that makes something more or bigger

+3

0 1 2 3 4 5 6 7 8 9 10

See also: **decrease**

The **increase** from 7 to 10 is 3.

independent variable
/in di PEN duhnt VAIR ee uh buhl/ (n.)

a variable with a value based on no other variable

$$x^2 - y = 1$$

The value of y depends on the value of x. The variable x is the **independent variable.**

See also: **dependent variable, variable**

---

inequality   /in i KWOL uh tee/ (n.)

a number sentence with quantities that are not the same

3 < 5   5 > 3

See also: **equal, greater than, less than**

Sara wrote the **inequality** 3 < 5.
An **inequality** uses the symbols <, >, or ≠.

---

infinity (∞)   /in FiN uh tee/ (n.)

a quantity that has no end or limit; a quantity so large it cannot be measured

1, 2, 3, ..., 267, ..., 912, ..., 10,340, ... ∞

The numbers go on to **infinity**.

---

input value
/IN put VAL yoo/ (n.)

a number that is entered into a formula

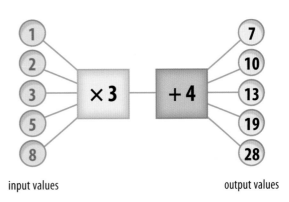

input values   output values

See also: **formula, output value**

An **input value** of 2 in the formula gives an output value of 10.

69

integer /IN tuh jur/ (n.)

a positive or negative whole number, or zero; a number that is in the set 0, 1, 2, 3… or in the set −1, −2, −3…

negative integers   positive integers

See also: **negative number, whole number, zero**

Every positive whole number is an **integer**. An example of a negative **integer** is −4.

interest /IN tur ist/ (n.)

the cost of borrowing money

interest

I borrowed $50. The **interest** was 10%. I paid back the $50 I borrowed, plus $5 **interest**.

interlocking cubes
/in tur LOK ing kyoobz/ (n.)

small cubes that connect to each other

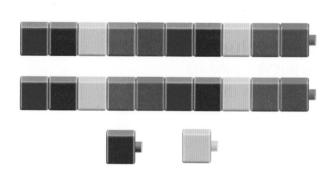

See also: **cube**

Make stacks of **interlocking cubes** to help you add. The cubes show that 11 plus 11 is 22.

intersect   /in tur SEKT/ (v.)
to cross; to share a point

See also: **cross, intersection**

The dots show where the lines and curves **intersect**.

---

intersection
/in tur SEK shuhn/ (n.)
the point where two or more
lines or curves meet

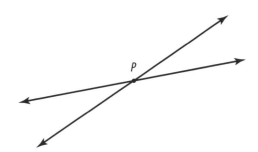

P

See also: **cross, intersect, point**

Point *P* is the **intersection** of the two lines.

---

interval   /IN tur vuhl/ (n.)
1. the distance between
   two numbers

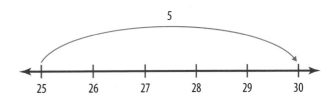

5

25   26   27   28   29   30

The **interval** between 25 and 30 is 5.

2. the time between two events

See also: **distance, time**

From 4:00 to 5:00 is an **interval** of one hour.

inverse /in VERS/ (n.)

an opposite

$$1 + -1 = 0$$

A number plus its additive **inverse** is 0.

$$5 \times \frac{1}{5} = 1$$

A number times its multiplicative **inverse** is 1.

---

inverse /in VERS/ (adj.)

opposite; reversed

$$7 \times 3 = 21$$

$$21 \div 3 = 7$$

Multiplication and division are **inverse** operations.

See also: **operation**

---

inverse proportion
/in VERS pruh POR shuhn/ (n.)

a relation where one quantity goes up as the other goes down

Rate or speed (r), in miles per hour	Time (t), in hours
10	6.0
20	3.0
30	2.0
60	1.0
120	0.5

See also: **area, direct proportion, proportion**

As the rate increases, the time decreases. The rate and time are an **inverse proportion**.

---

investigate /in VES tuh GAYT/ (v.)

to study a problem

I will **investigate** what happens when I add two odd numbers.

irregular /i REG yuh lur/ (adj.)

having sides that are two or more different lengths; having angles with two or more different measures

See also: **irregular polygon, regular polygon**

This pentagon is **irregular**. Its sides are not all the same length.

irregular polygon

/i REG yuh lur POL ee gon/ (n.)

a polygon with sides that are two or more different lengths

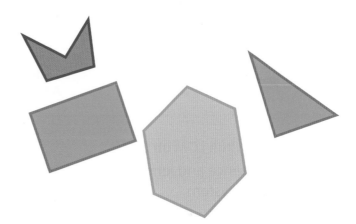

See also: **irregular, polygon, regular polygon**

A scalene triangle is an **irregular polygon**. Its sides are different lengths.

isosceles triangle

/ī SOS uh leez TRĪ ang guhl/ (n.)

a triangle with two equal sides

See also: **angle, side, triangle**

The two long sides of this **isosceles triangle** have equal lengths.

Kk

kilogram /KIL uh gram/ (n.)
a unit of mass in
the metric system

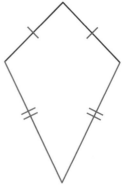

See also: **gram, metric system**

1 **kilogram** = 1,000 grams
A textbook might weigh about one **kilogram**.

kilometer /kuh LOM uh tur/ (n.)
a unit of distance and length
in the metric system

1 **kilometer** = 1,000 meters
A **kilometer** is a little more than half a mile.

See also: **meter, metric system**

kite /kīt/ (n.)
a quadrilateral with two pairs of
adjacent sides that are equal

See also: **adjacent, congruent, quadrilateral,
vertex**

The equal sides in a **kite** share a vertex. The two blue
sides in this **kite** are congruent, or equal. The two red
sides are also congruent.

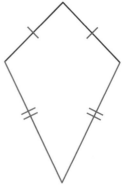

74

L /el/ (n.)
the roman numeral representing 50

The Roman numeral **L** is equal to 25 + 25.

See also: **Roman numerals**

label /LAY buhl/ (n.)
a word, phrase, or number
written beside an object
to describe it

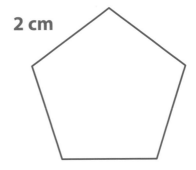
2 cm

The **label** shows the length of one of the figure's sides.

leap year /leep yeer/ (n.)
a year with 366 days

February has 29 days in a **leap year**.
Every fourth year is a **leap year**.

least common multiple
/leest KOM uhn MUL tuh puhl/ (n.)
the smallest whole number
that is a multiple of every
number in a set

The **least common multiple** of 6 and 4 is 12.
Multiples of 6: 6, 12, 18, 24
Multiples of 4: 4, 8, 12, 16, 20, 24

See also: **multiple**

length /length/ (n.)

1. the distance along a line or curve

The **length** of each figure is 5 centimeters.

2. one of the three dimensions of a three-dimensional object

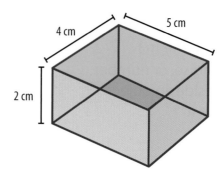

The **length** of the rectangular prism is 5 cm. Its width is 4 cm. Its height is 2 cm.

3. the amount of time an event lasts

The **length** of a day is 24 hours.

See also: **distance, three-dimensional, time, width**

---

less than (<) /les THan/ (adj.)

not as great as another number or amount

See also: **greater than**

$$7 \quad < \quad 10$$

Seven is **less than** ten. $7 < 10$

likely /LĪK lee/ (adj.)

having a good chance
of happening; probable

See also: **probable**

A bag has some counters. Most of the counters are
blue. You are **likely** to pull a blue counter from the bag.

line /līn/ (n.)

a straight object in geometry
that has no ends or thickness

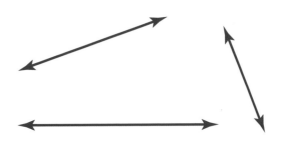

A **line** stretches on forever in both directions.

line of symmetry
/līn uhv SIM uh tree/ (n.)

a line that divides a shape
into congruent halves

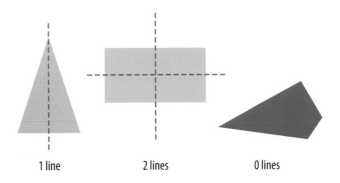

1 line   2 lines   0 lines

See also: **congruent, line, symmetrical, symmetry**

Some shapes have more than one **line of symmetry**.
Some shapes have no **line of symmetry**.

line plot /līn plot/ (n.)

a graph that uses Xs to show frequency in data

Number of Points Euna Scored

```
                                    X
                                    X
      X                    X        X
   ──────────────────────────────────────────
      15     16     17     18     19     20
```

See also: **data, frequency, number line**

The **line plot** shows Euna scored 19 points three times.

line segment /līn SEG muhnt/ (n.)

a part of a line described by its two endpoints

See also: **line, point**

Point C is one end of this **line segment**. Point D is the other end of the **line segment**.

list /list/ (n.)

a series of written names or items

Factors of 36
1
2
3
4
6
9
12
18
36

There is usually one item on each row of a **list**.

liter /LEE tur/ (n.)

a unit of volume
in the metric system

See also: **milliliter, volume**

1 **liter** = 1,000 milliliters
Water often comes in 1-**liter** bottles.

---

long /lawng/ (n.)

a base-ten block that
represents ten

See also: **cube, flat**

1 **long** = 10 cubes
You can show 12 with one **long** and two cubes.

---

long /lawng/ (adj.)

1. a description of distance,
   length, or time

   The string is 3 meters **long**.

   The movie is 2 hours **long**.

2. a great distance or length

   It is a **long** way from Mexico to China.

3. a large amount of time

   It took a **long** time to wash the dishes.

See also: **length, time**

Mm

M /em/ (n.)
the Roman numeral
that represents 1,000

M = 1,000
MI = 1,001
CM = 900

See also: **C, I, Roman numerals**

---

map /map/ (n.)
a picture that shows places
on Earth or in space

Stillwater
Tulsa
N
Oklahoma
City
Lawton

See also: **scale**

This **map** shows four cities in Oklahoma. Stillwater
is north of Oklahoma City.

---

mapping /MAP ing/ (n.)
a way of relating each member of
one set to a member of another

+2
2   9
5   4
7   11
9   7

See also: **set**

The **mapping** uses the rule *add* 2.

---

mass /mas/ (n.)
the amount of matter or material
in an object

See also: **weight**

We use grams and kilograms to measure **mass**.

maximum
/MAK suh muhm/ (adj.)

1. the greatest number
or amount in a set

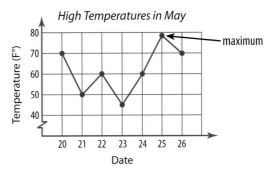

High Temperatures in May

The graph shows high temperatures in one week
of May. The **maximum** of this data set is 79°F.

2. the greatest number or amount
that is possible or allowed

The **maximum** prize is $50. You cannot
win more than that amount.

See also: **minimum, set**

---

mean /meen/ (n.)
the average of a set of numbers

3, 4, 5, 9, 9

$3 + 4 + 5 + 9 + 9 = 30$
$30 \div 5 = 6$
The **mean** is 6.

See also: **average, median, mode**

---

measure /MEZH ur/ (v.)
to find a size or quantity

See also: **length, measurement, meter**

Measure the length of a baseball bat with
a meter stick.

---

measurement /MEZH ur muhnt/ (n.)
a number and a unit that describes
a size or quantity

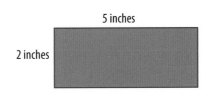

5 inches

2 inches

See also: **measure**

The rectangle's long side has a **measurement**
of 5 inches.

81

median /ME dee uhn/ (n.)

1. the middle value of a set of numbers when the numbers are in order

2. a line from a vertex of a triangle to the middle of the opposite side

See also: **average, mean, mode, triangle**

3, 3, **4**, 6, 9

The **median** is 4. It is the middle value in the set.

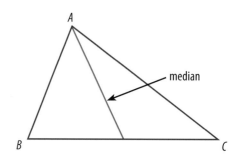

In triangle *ABC* the **median** goes from angle *A* to the middle of side $\overline{BC}$.

meter (m) /MEE tur/ (n.)

a unit of length or distance in the metric system

See also: **centimeter, kilometer, unit**

1 **meter** = 100 centimeters
1,000 **meters** = 1 kilometer
Doorknobs are about one **meter** above the floor.

meter stick /MEE tur stik/ (n.)

a measuring tool that is 100 centimeters long

See also: **measure, meter**

| 0 | 10 | 20 | 30 | 40 | 50 | 60 | 70 | 80 | 90 |

A **meter stick** is one meter long.

metric system
/MET rik SIS tuhm/ (n.)

a measurement system that is common around the world

See also: **customary system, measurement**

Basic Units in the Metric System	
Mass	gram
Length	meter
Capacity	liter

metric ton /MET rik tun/ (n.)
a unit of mass equal to one thousand kilograms

See also: **kilogram, mass, weight**

1 **metric ton** = 1,000 kilograms
An adult hippopotamus weighs about one **metric ton**.

mid- /mid/ (prefix)
a prefix used to describe the center of something

See also: **middle, midnight**

C is the **mid**point between A and B.

middle /MID uhl/ (n.)
the center; the part of a thing between its ends or sides

See also: **center**

Dani drew a dot in the **middle** of the circle.

midnight /MID nīt/ (n.)
12 o'clock at night

See also: **day, mid–, night**

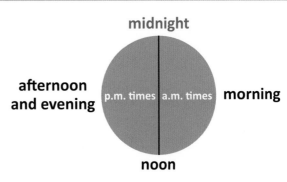

Midnight is 12 hours after noon.

mile (mi.) /mīl/ (n.)
a unit of long distance in the customary system

See also: **customary system, measure, unit**

1 **mile** = 5,280 feet

It takes about 15 minutes to walk a **mile**.

millennium /muh LEN ee uhm/ (n.)

a period of one thousand years

There was a **millennium** between the years 900 and 1900.

See also: **century**

---

milliliter (mL or ml)
/MIL uh lee tur/ (n.)

a unit of small volumes in the metric system

See also: **liter, measure, metric system, volume**

1,000 **milliliters** = 1 liter
A teaspoon holds about 5 **milliliters** of soup.

---

millimeter (mm)
/MIL uh mee tur/ (n.)

a unit of small lengths in the metric system

See also: **length, measure, meter, metric system**

1,000 **millimeters** = 1 meter
The cricket is about 25 **millimeters** long.

---

million /MIL yuhn/ (n.)

a number that is one thousand times one thousand

$$1,000,000 = 10^6$$

The number one **million** has six zeroes.
It can be written in different ways.

See also: **times, zero**

minimum /MIN uh muhm/ (adj.)

1. the least number or amount in a set

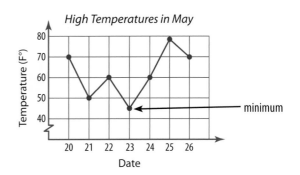

High Temperatures in May

The graph shows high temperatures in one week of May. The **minimum** of this data set is 45°F.

2. the least number or amount that is possible or allowed

The **minimum** age for the ride is 12. You must be at least 12 years old to go on the ride.

See also: **maximum**

---

minus (−) /mī nuhs/ (adj.)
negative; below zero

minus 5 degrees = 5 degrees below zero
It is **minus** 5 degrees Celsius, or −5° C, outside.

See also: **Celsius, Fahrenheit, negative number**

---

minus (−) /mī nuhs/ (prep.)
reduced or made less by a number; used to show subtraction

$$7 - 2 = 5$$

Seven **minus** two equals five.

See also: **subtract**

minute (min.) /MIN it/ (n.)

a unit of time equal to
sixty seconds

+ 1 minute

1 **minute** = 60 seconds
60 **minutes** = 1 hour

See also: **hour, minute hand, time**

minute hand /MIN it hand/ (n.)

the part of a clock that shows
the minute

minute hand

The clock shows 9:08. The **minute hand** points
to the eighth minute line.

See also: **clock, minute**

mixed number
/mikst NUM bur/ (n.)

a quantity with two parts;
the first part is a whole number
and the second part is a fraction

$$7\frac{5}{8}$$

whole number   fraction

$7\frac{5}{8}$ is a **mixed number**. The whole number is 7.
The fraction is $\frac{5}{8}$.

See also: **fraction, improper fraction, whole number**

mode /mohd/ (n.)

the value that appears most
often in a set of numbers

3, 3, 4, 6

3 appears most often. So, 3 is the **mode** of the set.

See also: **average, mean, median**

model /MOD uhl/ (n.)

a diagram or object that describes a mathematical idea

See also: **diagram**

The **model** shows a cone.

month /munth/ (n.)

a period of time equal to 28, 30, or 31 days

January

SUN	MON	TUE	WED	THU	FRI	SAT
				1	2	3
4	5	6	7	8	9	10
11	12	13	14	15	16	17
18	19	20	21	22	23	24
25	26	27	28	29	30	31

January is the first **month** of the year.
12 **months** = 1 year

See also: **day**

morning /MOR ning/ (n.)

the time of day between midnight and noon

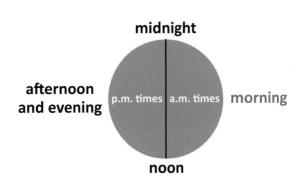

The sun rises each **morning**.

See also: **day, midnight**

multiple /MUL tuh puhl/ (n.)

a number formed by multiplying
a number by a whole number

Some Multiples of 5
1 x 5 = 5
2 x 5 = 10
3 x 5 = 15
4 x 5 = 20

A **multiple** of 5 is 15 because 3 times 5 is 15.

See also: **factor, multiply, product**

multiplication
/mul tuh pluh KAY shuhn/ (n.)

the act of finding the product
of two or more numbers

$$3 \times 4 = 4 + 4 + 4$$

An example of **multiplication** is $3 \times 4 = 12$.
Multiplication of whole numbers is repeated
addition.

See also: **addition, multiply, product**

multiply (× or ·)
/MUL tuh plī/ (v.)

to find the product of
two or more numbers

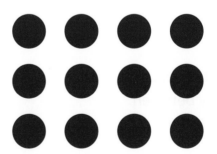

This array has 3 rows and 4 columns. It models
the multiplication fact 3×4.

If you **multiply** 3 by 4, you get 12.
It is the same as adding 3 to itself 4 times,
or $3 + 3 + 3 + 3$.

See also: **add, array, multiplication**

narrow /NAIR oh/ (adj.)
having edges that are close together; not wide

See also: **wide**

A ribbon is **narrow**. It is long, but it is not very wide.

---

negative number
/NE guh tiv NUHM bur/ (n.)
a number that is less than zero

See also: **integer, minus, positive number, zero**

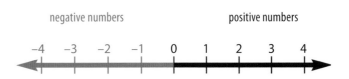

Use a minus sign to write a **negative number**.

---

net /net/ (n.)
a flat shape that can be folded to represent a solid shape

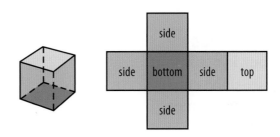

See also: **fold, solid**

The **net** of a cube has 6 squares.

---

night /nīt/ (n.)
the time from about 8:00 p.m. to 6:00 a.m.

See also: **day**

Most people sleep at **night**. It is usually dark at **night**.

nonagon /NON uh gon/ (n.)

a polygon with nine sides

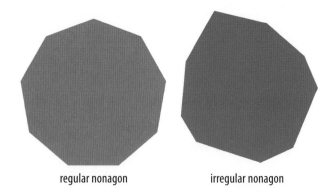

regular nonagon   irregular nonagon

All the sides of a regular **nonagon** are equal.
A **nonagon** has one more angle than an octagon.

See also: **angle, octagon, polygon, side**

number /NUHM bur/ (n.)

a symbol or a word used to count
or measure

The **number** 1 comes before the **number** 2.
The **number** 6.3 comes before the **number** 15.

See also: **decimal, number line, numeral**

number cubes
/NUHM bur kyoobz/ (n.)

small, six-sided cubes
often used in games

See also: **cube, probability, range**

Number cubes have different numbers or dots
on each side.

number fact /NUHM bur fakt/ (n.)

an equation that uses mostly
one-digit numbers

$5 + 4 = 9$ is a **number fact**. You can
use **number facts** to solve many
math problems.

See also: **equation, number**

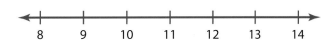

number line /NUHM bur līn/ (n.)

a line showing numbers in order

The **number line** shows whole numbers from 8 to 14.

See also: **compare, number, order**

---

number sentence
/NUHM bur SEN tuhns/ (n.)

an equation or an inequality

$6 \times 7 = 42$

$3 < 15$

$5 - 4 > 0$

See also: **algebra, equation, inequality**

A **number sentence** has the symbols =, <, or >.

---

numeral /NOO mur uhl/ (n.)

a symbol that represents a number

14 is a **numeral**.
The Roman **numeral** for 5 is V.

See also: **number, Roman numeral, symbol**

---

numerator /NOO muh ray tur/ (n.)

the top number of a fraction

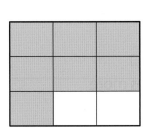

numerator

$$\frac{7}{9}$$

denominator

The **numerator** tells the number of parts that are being counted.

See also: **denominator, fraction**

Oo

oblique /oh BLEEK/ (adj.)
 slanted; not horizontal or vertical

W
The lines that form the letter W are all **oblique**.

H
The letter H has no **oblique** lines.

See also: **horizontal, line, vertical**

---

oblong /OB lawng/ (adj.)
 longer along one dimension than along the other

longer side

shorter side

Many rectangles are **oblong**.

See also: **rectangle, square**

---

obtuse angle
/uhb TOOS ANG guhl/ (n.)
 an angle that measures more than 90° and less than 180°

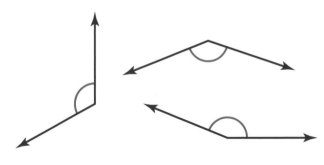

See also: **acute angle, angle, measure, obtuse triangle**

Each **obtuse angle** has a measure greater than 90°.

obtuse triangle
/uhb TOOS TRĪ ang guhl/ (n.)
  a triangle with one angle greater than 90°

See also: **acute triangle, obtuse angle, right triangle, triangle**

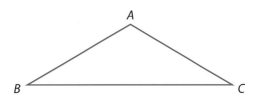

Angle *A* in this **obtuse triangle** measures 105°.

octagon /OK tuh gon/ (n.)
  a polygon with eight sides

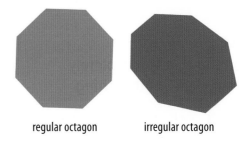

regular octagon   irregular octagon

See also: **angle, polygon, side**

All the sides of a regular **octagon** are equal. A stop sign is an example of an **octagon**.

octahedron
/ok tuh HEE druhn/ (n.)
  a solid with eight faces

See also: **face, polyhedron, solid**

An **octahedron** can look like two pyramids joined together. All the faces of a regular **octahedron** are triangles.

odd /od/ (adj.)
  not evenly divisible by two

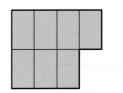

7

See also: **divisible, even, whole number**

Seven is an **odd** number. The opposite of **odd** is *even*.

operation /op uh RAY shuhn/ (n.)

addition, subtraction, multiplication, or division performed on a pair of numbers

$$8 + 2 = 10 \qquad 8 - 2 = 6$$
$$8 \times 2 = 16 \qquad 8 \div 2 = 4$$

Each **operation** has a different answer.

See also: **addition, division, multiplication, subtraction**

opposite /OP uh zit/ (n.)

1. something that is completely different from another thing

Day is the **opposite** of *night*.

2. a number that is the same distance from zero as another number

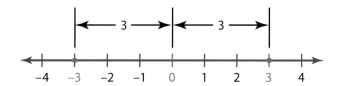

See also: **distance, minus**

The number 3 is the **opposite** of −3.

opposite /OP uh zit/ (prep.)

across from; on the other side of

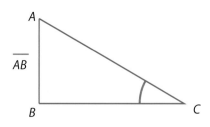

See also: **adjacent, angle, triangle**

Side *AB* of the triangle is **opposite** angle *C*.

order   /OR dur/ (v.)
to arrange in a sequence

You can **order** 5, 2, and 8 from least to greatest. Write them this way: 2, 5, 8.

See also: **compare, greater than, less than**

---

order   /OR dur/ (n.)
a sequence or arrangement

8, 9, 10, 11, 12, 13	in **order** from least to greatest
13, 12, 11, 10, 9, 8	in **order** from greatest to least

---

ordered pair   /OR durd pair/ (n.)
a set of numbers that describes a point on the coordinate grid

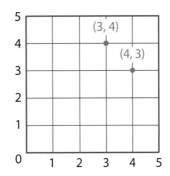

See also: **coordinate grid, coordinates, *x*-coordinate, *y*-coordinate**

The **ordered pair** (4, 3) is a different point from (3, 4). You write an **ordered pair** inside parentheses.

---

order of operations
/OR dur uhv op uh RAY shuhnz/ (n.)

a set of rules that shows which calculations to do first in an expression

$$10 \times (5 - 1) + 3^2$$

1. Parentheses	$10 \times 4 + 3^2$
2. Exponents	$10 \times 4 + 9$
3. Multiply and/or divide.	$40 + 9$
4. Add and/or subtract.	49

See also: **operation**

The **order of operations** says to multiply before you add.

ordinal /ORD nuhl/ (adj.)

having to do with order
or position

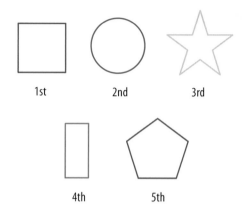

1st   2nd   3rd

4th   5th

See also: **order, second, third**

1st, 2nd, 3rd, 4th, and 5th are **ordinal** numbers. You can use an **ordinal** number to tell where something is in line.

---

origin /OR uh jin/ (n.)

the place on a graph where
the two axes cross

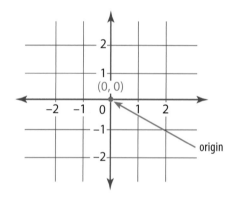

See also: **axis, coordinate grid, graph, ordered pair**

The ordered pair (0, 0) describes the **origin**.
The point (0, 2) is two spaces above the **origin**.

---

ounce (oz.) /owns/ (n.)

a unit of weight in the
customary system

See also: **customary system, pound, ton, weight**

A slice of bread weighs about one **ounce**.
Sixteen **ounces** equal one pound.

outcome /OWT kuhm/ (n.)
the result of an experiment

See also: **event, number cubes, probability**

Pierre tosses a six-sided number cube. The **outcome** is 5. There are six possible **outcomes**.

output value
/OWT put VAL yoo/ (n.)
a number that is determined by a formula

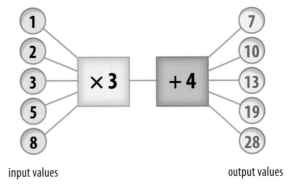

input values   output values

See also: **formula, input value, rule**

The formula is "multiply by 3 and add 4." If the input value is 2, the **output value** is 10.

oval /OH vuhl/ (n.)
1. ellipse

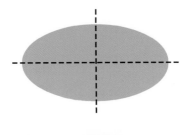

An **oval** has two lines of symmetry.

2. a figure shaped like an egg

An **oval** can be narrow on one end and wider on the other.

See also: **circle, ellipse, line of symmetry**

pair /pair/ (n.)

1. a set of two things that belong together

Olivia bought a new **pair** of shoes.

2. a set of two things that can be put together

Ted and Gina work together. They work as a **pair**.

parallel (||) /PAIR uh lel/ (adj.)

describing lines that never meet

See also: **cross, perpendicular**

The lines in the picture are **parallel**. **Parallel** lines never cross.

parallelogram
/pair uh LEL uh gram/ (n.)

a four-sided figure with two pairs of parallel sides

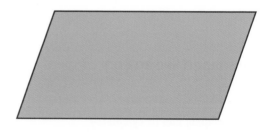

See also: **pair, parallel, quadrilateral, side**

A **parallelogram** has opposite sides that are parallel.

parentheses [( )]
/puh REN thuh seez/ (n.)

symbols that show what
operation to do first

parentheses

$$(6 + 4) \div 2$$

Tulia solves the expression $(6 + 4) \div 2$.
The **parentheses** tell her to add 6 and 4 first.

See also: **brackets, order of operations, pair**

part /part/ (n.)

a section or piece of an object
or a group

part

The yellow section is a **part** of the rectangle.

See also: **fraction, whole**

pattern /PAT urn/ (n.)

a sequence that follows a rule

1, 3, 5, 7, 9, 11...

This **pattern** is all odd numbers. It follows
the rule "add 2."

See also: **odd, order, rule, sequence**

pay /pay/ (v.)

to give money in exchange
for a thing or a service

Padma must **pay** $5 for the hat.

Mr. Ruiz will **pay** Joe $10 to wash his car.

See also: **cost, exchange**

pentagon /PEN tuh gon/ (n.)
a polygon with five sides

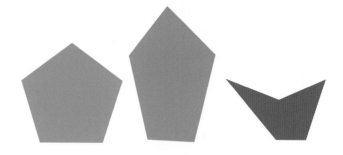

See also: **angle, polygon, side**

All the angles of a regular **pentagon** are equal.
A **pentagon** has one more side than a square.

pentomino /pen TOM uh noh/ (n.)
a shape made from five
identical squares

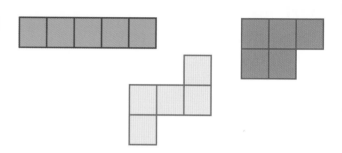

A **pentomino** is a flat shape. The sides of the squares
in a **pentomino** are connected.

per /pur/ (prep.)
for each; in each

Nina drives 100 km in one hour.
Her speed is 100 km **per** hour.

Each student has 2 pencils.
So, there are 2 pencils **per** student.

See also: **kilometer, rate, speed**

percent (%) /pur SENT/ (adv.)
out of one hundred;
per one hundred

$50\% = \frac{50}{100} = 0.5$

$25\% = \frac{25}{100} = 0.25$

Fifty **percent** means 50 out of 100.
$\frac{25}{100}$ is equal to 25 **percent**.

See also: **decimal, fraction, per**

perfect square
/PUR fikt skwair/ (n.)

an integer that is the product of a number and itself

$9 \times 9 = 81$

$12 \times 12 = 144$

81 is a **perfect square**.
144 is also a **perfect square**.

See also: **integer, square, squared, square root**

---

perimeter /puh RI muh tur/ (n.)

the distance around a flat shape

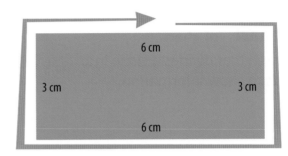

See also: **area, circumference, measurement**

The rectangle's **perimeter** is 18 cm.

---

perpendicular (⊥)
/pur puhn DIK yuh lur/ (adj.)

intersecting at a right angle

See also: **cross, intersect, parallel, right angle**

Perpendicular lines meet at 90°. The letter L is made of two **perpendicular** line segments.

---

pi (π) /pī/ (n.)

the ratio of the circumference to the diameter of a circle

$\pi \approx 3.14$

The value of **pi** is a little more than 3.

See also: **circle, circumference, diameter, radius, ratio**

pictogram /PIK tuh gram/ (n.)

a graph that uses pictures
to stand for quantities

Favorite Seasons					
Spring	☺	☺	☺		
Summer	☺	☺	☺	☺	☺
Fall	☺	☺			
Winter	☺				

Key: ☺ = 2 students

See also: **graph**

This **pictogram** shows students' favorite seasons.
The key tells you each face stands for two students.

pie chart /pī chart/ (n.)

a graph that divides a circle into
parts to show information

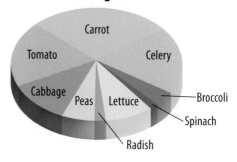

Favorite Vegetables

See also: **chart, circle graph, graph**

The **pie chart** shows students' favorite vegetables.
The red part of this **pie chart** is very small.
Few students chose radishes.

pint (pt.) /pīnt/ (n.)

a unit used to measure capacity
in the customary system

A **pint** is a little less than a liter.
A glass holds about half a **pint** of water.

See also: **capacity, cup,
customary system, liter**

place /plays/ (n.)

the position of a digit
in a numeral

hundreds tens ones

752

See also: **digit, numeral, place value**

The digit 2 is in the ones' **place**.
The digit 7 is in the hundreds' **place**.

place value /plays VAL yoo/ (n.)

the value of a digit in a numeral

4,356

Use **place value** to read the number 4,356. The 4 means "4 thousands." The 3 means "3 hundreds." The 5 means "5 tens." The 6 means "6 ones."

See also: **digit, numeral, place**

plane /playn/ (n.)

a flat surface that goes on forever in all directions

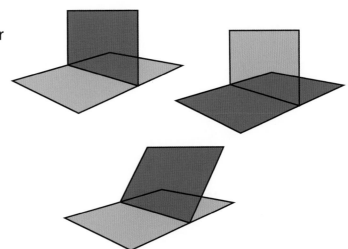

See also: **flat, line, two-dimensional**

A **plane** has no thickness. Each picture shows two **planes** intersecting.

plane of symmetry

/playn uhv SI muh tree/ (n.)

a plane dividing a solid shape into halves that reflect each other

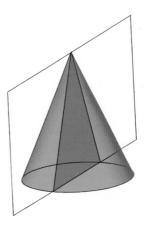

See also: **line of symmetry, plane, symmetrical**

You can draw a **plane of symmetry** on a cone.

103

plane shape /playn shayp/ (n.)
a flat figure; a two-dimensional object

See also: **dimension, flat, plane, two-dimensional**

A **plane shape** has length and width. It does not have thickness.

plot /plot/ (v.)
to mark a point or group of points on a graph

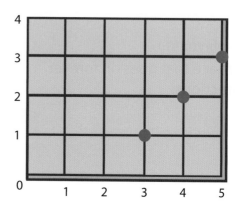

See also: **graph, line, point**

Plot the points (3, 1), (4, 2), and (5, 3). The points form a line.

plus (+) /pluhs/ (prep.)
increased or made more by a certain number

$$8 + 5 = 13$$

Eight **plus** five equals thirteen.

See also: **add, minus**

p.m. /pee em/ (abbr.)
after 12:00 noon and before 12:00 midnight

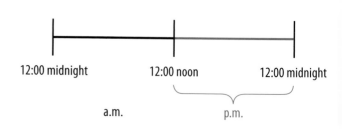

See also: **a.m., midnight**

The time is 3:00 in the afternoon. We say it is 3:00 **p.m.**

ABCDEFGHIJKLMNO**P**QRSTUVWXYZ   **polyhedron**

point /point/ (n.)

an object with no dimensions
in geometry

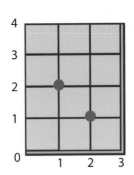

See also: **dimension, geometry,
graph, represent**

A dot can represent a **point** on a graph.

polygon /POL ee gon/ (n.)

a closed plane shape
with no curves

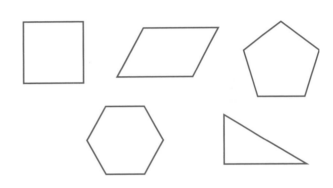

See also: **plane shape**

Each shape is a kind of **polygon**.

polyhedron /pol ee HEE druhn/ (n.)

a solid shape with no
curved surfaces

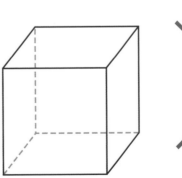
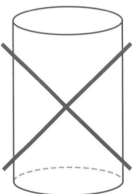

See also: **face, polygon**

A cube is a **polyhedron**. A cylinder is not a **polyhedron**.
Every face of a **polyhedron** is a polygon.

position /puh ZI shuhn/ (n.)

location of an object compared
to other objects

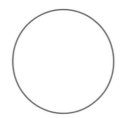

The circle is to the left of the square. Its **position** is
on the left.

positive number
/POZ uh tiv NUHM bur/ (n.)

a number greater than 0

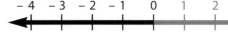

negative numbers   positive numbers

$$-4 \quad -3 \quad -2 \quad -1 \quad 0 \quad 1 \quad 2 \quad 3 \quad 4$$

See also: **greater than, integer,
negative number, zero**

4 is greater than 0. It is a **positive number**.

pound (lb.) /pownd/ (n.)

a unit for measuring weight
in the customary system

One **pound** is equal to 16 ounces.
A loaf of bread weighs about one **pound**.

See also: **customary system, measure,
ounce, weight**

power /POW ur/ (n.)

the number of times a number
is multiplied by itself

power

$$5^3 = 5 \times 5 \times 5$$

Five to the third **power** is written 5^3, or $5 \times 5 \times 5$.

See also: **exponent, multiplication**

predict /pri DIKT/ (v.)

to guess what will happen

I will toss two number cubes.
I **predict** that their sum will be seven.

price /prīs/ (n.)

the cost of an item or service

See also: **cost, pay**

The ball's **price** is $3.50.

---

prime factor /prīm FAK tur/ (n.)

a factor that is divisible
only by itself and 1

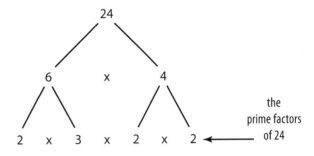

the prime factors of 24

See also: **factor, prime number**

Two is a factor of 24. Two is divisible only by itself and one. Two is a **prime factor** of 24.

---

prime number
/prīm NUHM bur/ (n.)

a whole number with exactly
two factors: itself and 1

17

17 has just two factors: 17 and 1.
So, 17 is a **prime number.**

See also: **factor, whole number**

---

prism /PRI zuhm/ (n.)

a polyhedron with the same kind
of polygon at each end

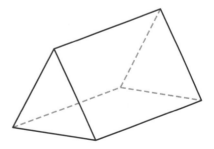

See also: **face, rectangle**

A triangular **prism** has 5 faces. Two faces are triangles.
The rest are rectangles.

probability
/prob uh BIL uh tee/ (n.)

the chance that an event
will happen

See also: **event, likely, outcome, probable**

There is one chance in two that a coin will land on tails.
The **probability** of getting tails is $\frac{1}{2}$.

probable /PROB uh buhl/ (adj.)

having a good chance
of happening; likely

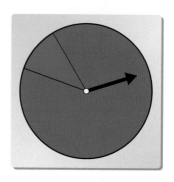

See also: **likely, probability**

Most of the spinner is blue. It is **probable** that
the arrow will point to a blue area of the spinner.

product /PROD uhkt/ (n.)

the answer to a multiplication
problem

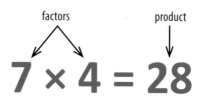

$$7 \times 4 = 28$$

Multiplying two factors gives you their **product**.
The **product** of 7 and 4 is 28.

See also: **answer, factor, multiplication,
quotient**

profit /PROF it/ (n.)

the difference between the
buying price and selling price
of an item

See also: **cost, pay, price**

Roshni bought a book for $3. She sold it for $5.
She made a **profit** of $2.

proper fraction
/PROP ur FRAK shuhn/ (n.)

 a fraction with a value
 less than 1

See also: **improper fraction**

An example of a **proper fraction** is $\frac{1}{2}$.
It is less than 1.

An example of an improper fraction is $\frac{4}{3}$.
It is greater than 1.

---

property /PROP ur tee/ (n.)
 a characteristic of an object
 or number

One **property** of a square is
that it has four angles.

---

proportion /pruh POR shuhn/ (n.)
 1. a statement that two ratios
   are equal

10 divided by 2 is 5.
15 divided by 3 is also 5.
The ratios are the same.
$\frac{2}{10} = \frac{3}{15}$ is a **proportion**.

 2. a relation where the sizes of
   objects change in the same way

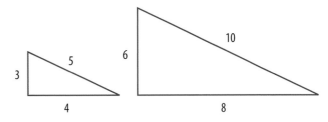

See also: **equal, equivalent, ratio, triangle**

The small triangle has sides of 3, 4, and 5 units.
The large triangle has sides of 6, 8, and 10 units.
The triangles are in **proportion**.

---

pyramid /PEER uh mid/ (n.)
 a polyhedron in which all
 the faces except one meet
 at a single point

See also: **base, face, polyhedron,
square, triangle**

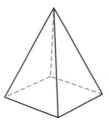

One face of a **pyramid** is called the base.
All the other faces are triangles.

Qq

quadrant /KWAHD ruhnt/ (n.)

one of the four areas
of a coordinate grid

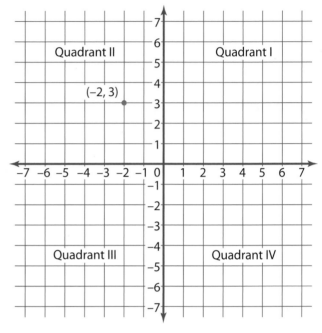

The point (−2, 3) is in the second **quadrant**.

See also: **coordinate grid, coodinates, x-axis, y-axis**

quadrilateral
/kwahd ruh LA tuh ruhl/ (n.)

a polygon with four sides

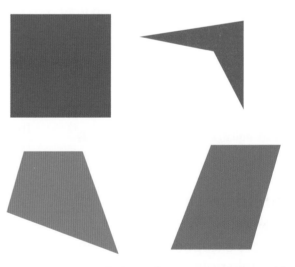

A square is a **quadrilateral**. Every **quadrilateral** has four angles.

See also: **angle, polygon, side, square**

quart /kwort/ (n.)

a unit for measuring capacity
in the customary system

> One **quart** is equal to 2 pints.
> Four **quarts** make one gallon.
> A **quart** is a little more than one liter.

See also: **capacity, customary system, gallon, liter, pint**

quarter /KWOR tur/ (n.)

one of four equal parts;
one fourth

See also: **fraction, half**

One **quarter** of the circle is light blue.

questionnaire
/kwes chuh NAIR/ (n.)

a list of questions used
to collect data

> Questionnaire
> by Leo
>
> 1. Do you play sports?
>   Yes   No
>
> 2. Do you like to sing?
>   Yes   No

Leo wrote a
questionnaire
with 2 questions.

See also: **data**

quotient /KWOH shuhnt/ (n.)

the answer to a division problem

$$30 \div 3 = 10 \quad \longleftrightarrow \quad \frac{10}{3)30}$$

See also: **answer, divide, product**

Divide 30 by 3. The **quotient** is 10.

111

Rr

radius (r) (plural radii)
/RAY dee uhs/ (n.)

a line segment from the center of a circle to its outer edge

See also: **circle, diameter**

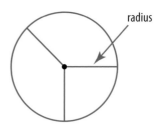
radius

Each of the lines in the picture is a **radius**.
The **radius** of a circle is half its diameter.

---

random /RAN duhm/ (adj.)

by chance; without being planned

A coin will land on heads or tails when it is tossed. You cannot choose which one you will get. The result is **random**.

See also: **event, probability, trial**

---

range /raynj/ (n.)

the difference between the greatest and least numbers in a set

2, 4, 5, 9, 10, 12

The greatest number in this set is 12.

The least number is 2.

$12 - 2 = 10$

The **range** of the set is 10.

See also: **difference, set**

---

rate /rayt/ (n.)

a comparison of two quantities that are expressed in different units

Lupe earns $40 in two hours.
Her **rate** of pay is $20 per hour.

Andre runs 25 meters in five seconds.
His **rate** of speed is 5 meters per second.

See also: **per, ratio, unit**

ratio /RAY shee oh/ (n.)

a comparison of two quantities

See also: **fraction, proportion, rate**

There are 3 circles and 2 squares. The **ratio** of circles to squares is 3:2, also written as $\frac{3}{2}$.

---

rational number
/RASH uh nuhl NUHM bur/ (n.)

a number that can be written as a fraction with integers in the numerator and denominator

See also: **denominator, fraction, integer, numerator, quotient**

$$0.5 = \frac{1}{2}$$

You can write 0.5 as $\frac{1}{2}$. So, 0.5 is a **rational number**.

---

ray /ray/ (n.)

a part of a line with exactly one endpoint

See also: **angle, endpoint, line**

One end of a **ray** goes on forever. The sides of this angle are formed by **rays**.

---

reasonable /REE znuh buhl/ (adj.)
sensible; logical

Tariq says that 21 × 41 is a little more than 800.
20 × 40 = 800
Tariq's answer is **reasonable**.

---

record /ri KORD/ (v.)
to write down data

See also: **chart, data**

What is your favorite sport?

Basketball |||

Baseball ||

Soccer ||||

Jun and Bill ask their friends some questions. They **record** the answers.

rectangle /REK tan guhl/ (n.)

a quadrilateral with four
right angles

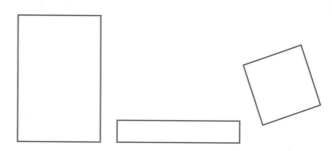

See also: **quadrilateral, right angle, square**

The opposite sides of a **rectangle** are equal.
Every square is a **rectangle**. Not every **rectangle**
is a square.

reduce /ri DOOS/ (v.)

1. to make less or smaller

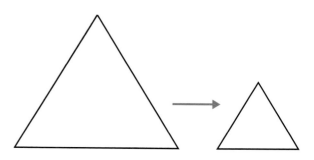

You can **reduce** the triangle on a copy machine.

2. to express a fraction
   in least terms

$$\frac{3}{6} \longrightarrow \frac{1}{2}$$

See also: **fraction, increase, simplify**

You can **reduce** $\frac{3}{6}$ to $\frac{1}{2}$. The value stays the same.

reflect /ri FlEKT/ (v.)

to create a mirror image
of an object

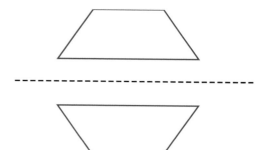

See also: **reflection, rotate, slide**

You can **reflect** a shape by flipping it over.
A shape might point in the other direction
after you **reflect** it.

reflection /ri FlEK shuhn/ (n.)

a mirror image

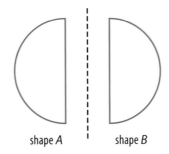

shape *A*   shape *B*

Shape *A* is a **reflection** of Shape *B*.

---

reflective symmetry

/ri **FLEK** tiv **SI** muh tree/ (n.)

the property of having two halves that are mirror images

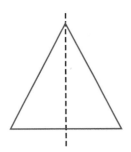

The triangle in the picture has **reflective symmetry**. The line of symmetry is shown.

---

reflex angle

/**REE** fleks **ANG** guhl/ (n.)

an angle measuring more than 180° and less than 360°

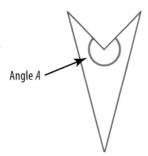

Angle *A*

Angle *A* measures 270°. So, it is a **reflex angle**.

---

regroup /ree GROOP/ (v.)

to reorganize the place value of a number and use it to simplify your work

$$\begin{array}{r} 1 \\ 38 \\ +\ 15 \\ \hline 3 \end{array}$$

Add 38 and 15. There are 13 ones. **Regroup** them to make 1 ten and 3 ones. You often need to **regroup** when you add or subtract.

regular polygon
/REG yuh lur POL ee gon/ (n.)

a polygon with sides all of
equal length

See also: **angle, polygon, regular polyhedron, side**

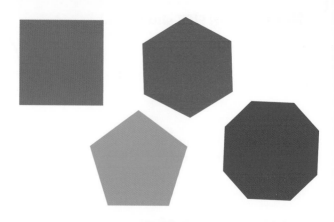

Each shape is a **regular polygon**.

regular polyhedron
/REG yuh lur pol ee HEE druhn/ (n.)

a polyhedron with faces that are
all the same

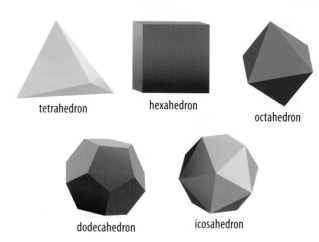

tetrahedron

hexahedron

octahedron

dodecahedron

icosahedron

See also: **congruent, face, polyhedron, regular polygon**

A cube is a **regular polyhedron**. It has six equal faces.
The faces of a **regular polyhedron** meet at the
same angle.

remainder (R) /ri MAYN dur/ (n.)
the amount left after solving
a division problem

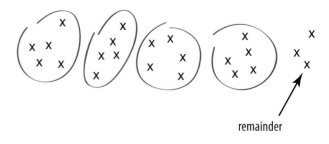

remainder

See also: **divide, division, quotient**

Divide 23 by 5. There are 3 left over.
The **remainder** is 3.

116

repeat /ri PEET/ (v.)

1. to happen again

0.452452452452...

The digits 452 **repeat** again and again.

2. to do something again in the same way

25, 30, 35, 40, 45, 50, ...
    +5   +5   +5   +5   +5

Ken starts with 25. He adds 5 again and again. He can **repeat** this step many times.

repeating decimal
/ri PEE ting DES uh muhl/ (n.)

a decimal with digits that appear in the same order forever

$$\frac{1}{9} = 0.11111... = 0.\overline{1}$$

$$\frac{4}{7} = 0.571428571428... = 0.\overline{571428}$$

See also: **decimal, fraction, rational number, repeat**

You can express a **repeating decimal** as a fraction. Draw a line above the digits that repeat.

represent /rep ri ZENT/ (v.)

to stand for; to be equivalent to; to take the place of

In 642 the digit 6 **represents** 600.

revolve /ri VOLV/ (v.)

to move in a path around another object

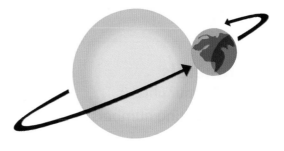

See also: **ellipse, rotate**

Earth **revolves** around the sun.

rhombus /ROM buhs/ (n.)

a quadrilateral with four equal sides

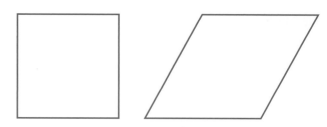

See also: **parallelogram, quadrilateral, side, square**

Every square is also a **rhombus**. Every **rhombus** is also a parallelogram.

right angle (∟) /rīt ANG guhl/ (n.)

an angle measuring 90 degrees

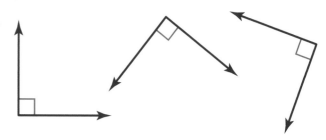

See also: **acute angle, degree, obtuse angle, perpendicular, right triangle**

Two perpendicular lines form a **right angle**. A **right angle** is one fourth of a complete turn.

right triangle
/rīt TRĪ ang guhl/ (n.)

a triangle with one 90-degree angle

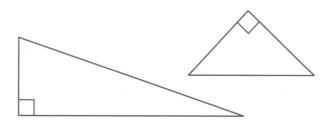

See also: **right angle, triangle**

Each triangle has a right angle. So, each is a **right triangle**.

Roman numerals
/ROH muhn NOO mur uhlz/ (n.)

the system of writing numbers used in ancient Rome

I = 1   V = 5   X = 10

Roman numerals use letters that represent quantities. To show 16 in **Roman numerals**, write XVI. **Roman numerals** have no symbol for zero.

See also: **I, numeral, represent, V, X**

rotate /ROH tayt/ (v.)

to turn or to spin, especially around an axis

See also: **axis, rotation**

Spin the globe. It will **rotate** around its axis.

rotation /roh TAY shuhn/ (n.)

one complete turn around an axis or point

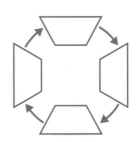

See also: **axis, reflection, rotate, transformation, translation**

The picture shows a trapezoid making one full **rotation**.

rotational symmetry
/roh TAY shuhn uhl SI muh tree/ (n.)

the property of looking the same when turned less than 360°

An equilateral triangle fits into its outline 3 times in one rotation.

This shape fits into its outline 4 times in one rotation.

This letter fits into its outline twice in one rotation.

See also: **line of symmetry, rotate, rotation, symmetry**

Rotate the triangle 120 or 240 degrees. It looks exactly the same. The triangle has **rotational symmetry**.

round /rownd/ (v.)

to approximate the value of a number by using a particular place value

Round 6,241 to the nearest hundred.
6,241 is closer to 6,200 than it is to 6,300.
Round 6,241 down to 6,200.

See also: **approximate, power**

row /roh/ (n.)

1. numbers or information set horizontally

The middle **row** shows the numbers from 5 to 8.

2. a group of objects placed side by side

Derrick planted a **row** of flowers.

See also: **chart, column, horizontal, table**

---

rule /rool/ (n.)

1. a set of steps used to find an answer

8 cm

I want to find the perimeter of the square. I can use the **rule** "multiply the length of one side by 4."

2. an instruction for creating a pattern

7, 14, 28, 56, 112, 224, ...
×2   ×2   ×2   ×2   ×2   ×2

The **rule** for this pattern is "multiply by 2."

See also: **answer, formula, pattern, perimeter**

---

ruler /ROO lur/ (n.)

a tool used to measure length

See also: **length, measure, measurement**

A **ruler** has a straight edge. You can use a **ruler** to measure in centimeters or inches.

scale /skayl/ (n.)

1. a tool used to measure weight or mass

Al stepped on the **scale**. He weighed 125 pounds.

2. a system of units used to measure

You can measure temperature on the Celsius **scale**. Write the temperature in degrees Celsius.

3. a ratio that compares distances on a map to distances in the real world

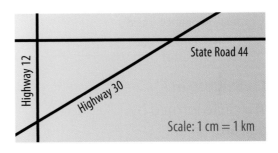

Highway 12

Highway 30

State Road 44

Scale: 1 cm = 1 km

One cm on the map stands for 1 km in real life. The map's **scale** is 1:100,000.

See also: **map, measure, ratio, unit**

scale drawing

/skayl DRAW ing/ (n.)

a drawing of objects that are in proportion to real objects

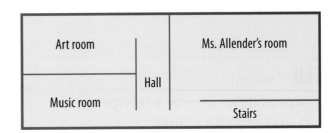

Art room

Ms. Allender's room

Hall

Music room

Stairs

Abdul makes a **scale drawing** of his school. His **scale drawing** is much smaller than the real school.

See also: **diagram, proportion, scale**

scalene triangle
/SKAY leen TRĪ ang guhl/ (n.)
   a triangle with no equal sides

See also: **acute triangle, angle, side, triangle**

The lengths of the sides of each triangle are all different. Each triangle is a **scalene triangle**.

second (sec.) /SE kuhnd/ (n.)
   a unit used to measure time

60 **seconds** = 1 minute
Measure short amounts of time in **seconds**.

See also: **hour, minute, time**

second /SE kuhnd/ (adj.)
   after the first in a sequence

8, 16, 32, 64, 128, …

The **second** number in the sequence is 16.

See also: **ordinal, sequence, third**

sector /SEK tur/ (n.)
   part of a circle formed by
   two radii and an arc

See also: **arc, circle, quadrant, radius**

The red part of the circle is a **sector**.

sell /sel/ (v.)
   to provide an item or a service
   in exchange for money

The store owner will **sell** this doll for $10.
The opposite of *sell* is buy.

See also: **cost, opposite, pay, price**

semicircle /SE mee sur kuhl/ (n.)
half of a circle

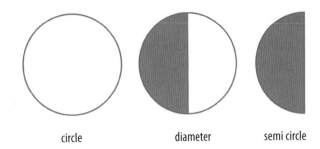

circle   diameter   semi circle

See also: **circle, diameter**

The straight side of a **semicircle** is a diameter.

sequence /SEE kwents/ (n.)
a set of numbers in
a particular order

$$8, 4, 2, 1, \frac{1}{2}, \frac{1}{4}, \ldots$$

A **sequence** is often a pattern. The rule
for this **sequence** is "divide by 2."

See also: **pattern, rule, set**

set /set/ (n.)
a group of numbers or objects

See also: **group, triangle**

All the shapes in this **set** are triangles.

shade /shayd/ (v.)
to color or to make darker

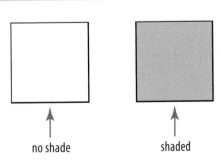

no shade   shaded

Shade the box on the right. Do not **shade** the box
on the left.

share /shair/ (v.)

to divide an object or group of objects into parts

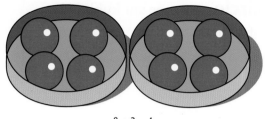

$$8 \div 2 = 4$$

See also: **divide, equal**

Tony and Maria will **share** eight marbles. They will each get four marbles.

short /short/ (adj.)

1. small in length or height

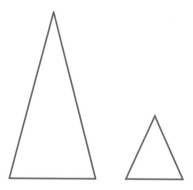

The triangle on the left is tall. The triangle on the right is **short**.

2. for a small amount of time

We ran around the track. Then we took a **short** rest.

See also: **long**

side /sīd/ (n.)

a line segment that is one part of a polygon

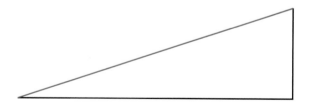

See also: **curve, line segment, polygon**

A triangle has three **sides**. The red line is one **side** of a triangle.

sign /sīn/ (n.)

a symbol used in a number sentence

See also: **compare, operation, symbol**

plus sign
↓
$$8 + 2 = 10$$

times sign
↓
$$36 = 6 \times 6$$

greater than sign   minus sign
↓   ↓
$$58 > 63 - 7$$

You can use a **sign** to show an operation or to compare. The symbol "=" is called an equals **sign**.

---

simplify /SIM pluh fī/ (v.)

to rewrite an expression using shorter terms or smaller numbers that have the same value

See also: **equivalent, fraction, square root, value**

$$\frac{12}{16} = \frac{3}{4}$$

You can **simplify** $\frac{12}{16}$ to $\frac{3}{4}$.

$$\sqrt{81} = 9$$

You can **simplify** $\sqrt{81}$ by rewriting it as 9.

---

size /sīz/ (n.)

the dimensions of an object

See also: **dimension, length, volume, width**

An object's **size** can be its length, its weight, or its volume.

---

slide /slīd/ (v.)

to move a shape without turning or flipping it

See also: **flip, translate**

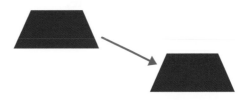

Roy will **slide** the trapezoid down and to the right.

slow /sloh/ (adj.)
moving at a low speed

Slow is the opposite of fast.

See also: **opposite, speed**

---

solid /SOL id/ (n.)
an object that has
three dimensions

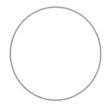

A sphere is a **solid**. It has height, length, and width.
A circle is not a **solid**. It has length and width but
not height.

See also: **dimension, height, length, width**

---

solution /suh LOO shuhn/ (n.)
the answer to a problem

$$(5 + 3) \times 10 = \underline{\quad}$$

The problem is $(5 + 3) \times 10 = \underline{\quad}$.
The **solution** is 80.

See also: **answer, solve**

---

solve /sawlv/ (v.)
to find the answer to a problem

Solve for x.
$$x - 15 = 5$$
$$x - 15 + 15 = 5 + 15$$
$$x = 20$$

See also: **algebra, solution, variable**

---

speed /speed/ (n.)
distance traveled in a unit of time,
such as miles per hour

See also: **distance, rate, time, unit**

Mr. Qan drove at a **speed** of 60 miles per hour.

spend /spend/ (v.)
to pay money

We plan to **spend** about $50 at the store.

See also: **cost, pay, price**

---

sphere /sfeer/ (n.)
a solid object that is completely round

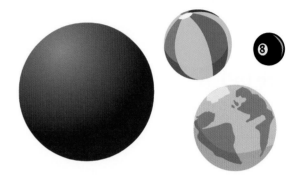

See also: **circle, round, solid, spherical**

A **sphere** has no faces or edges.

---

spherical /SFEER uh kuhl/ (adj.)
in the shape of a sphere

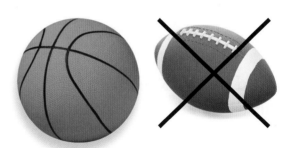

See also: **circle, round, solid, sphere**

The basketball is **spherical**. The football is not **spherical**.

---

spiral /SPĪ ruhl/ (n.)
a curve that rotates outward from a central point

See also: **curve, point, rotate**

The shell of a snail looks like a **spiral**.

square /skwair/ (n.)

a quadrilateral with four equal sides and four right angles

See also: **quadrilateral, rectangle, right angle, side**

A **square** is a type of rectangle.

---

square (n^2) /skwair/ (v.)

to multiply a number by itself

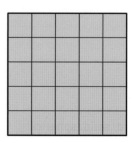

5 rows and 5 columns. $5 \times 5 = 25 = 5^2$

See also: **squared, square number, square root**

Multiply five by five to **square** five. Write 5^2.
$5^2 = 25$

---

squared (n^2) /skwaird/ (adj.)

multiplied by itself

$$10 \times 10 = 10^2$$
$$= 100$$

Ten **squared** can be written as 10^2.

See also: **square, square number, square root**

---

square number
/skwair NUHM bur/ (n.)

the product of a number and itself

$$6 \times 6 = 36$$

36 is a **square number**.

See also: **square, squared, square root**

square root (√) /skwair root/
(n.)

a number that can be multiplied by itself to get another number

$$\sqrt{9} = 3$$

$3 \times 3 = 9$. Three is the **square root** of nine.

See also: **square, squared, square number**

---

square unit /skwair YOO nit/
(n.)

a measure of area

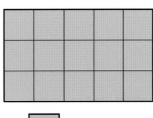

= 1 square unit

A square with sides of one unit measures one **square unit**. The floor measures 15 **square units**.

See also: **area, measure, square, unit**

---

standard form
/STAN durd form/ (n.)

the way numbers are usually written; numeral form

$$32 = 30 + 2$$
$$= 3 \times 10 + 2 \times 1$$

The number 32 is written in **standard form**. It means 3 tens and 2 ones.

See also: **digit, expanded form, numeral, place value**

standard unit
/STAN durd YOO nit/ (n.)

a measuring unit that is commonly used and has a definite value

Some Standard Customary Units
foot
pound
degree Fahrenheit
quart

Some Standard Metric Units
meter
kilogram
degree Celsius
liter

See also: **measurement, unit**

A mile is a **standard unit**. Everyone agrees about the length of a mile.

---

step /step/ (n.)

1. the distance between two numbers
2. one part of the process of solving a problem

There is a **step** of 5 from 23 to 28.

Solve $8 + (4 - 3) \times 5$

First **step**: Find $4 - 3$.
Second **step**: Multiply the difference by 5.
Third **step**: Add 8 to the product.

See also: **difference, order of operations, pattern, product**

---

straight angle
/strayt ANG guhl/ (n.)

an angle that measures 180 degrees

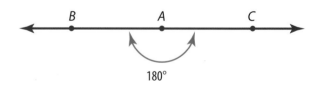

See also: **angle, line, line segment**

$\overline{AB}$ and $\overline{AC}$ form a **straight angle**. Angle *BAC* measures 180 degrees. A **straight angle** looks like a line or a line segment.

subset /SUHB set/ (n.)

a particular part of a set

1	5	13		29	65
4	12	28		791	993
36	454				
even numbers				whole numbers	

See also: **part, set, whole number**

Some whole numbers are even. Even numbers are a **subset** of the set of whole numbers.

subtract (−) /suhb TRAKT/ (v.)

1. to reduce one number by the value of another number

$$10 − 6 = 4$$

I have 10 dollars. I spend six dollars.
I **subtract** six from ten to find what is left.

2. to find the difference between two numbers

Six marbles are in the top row. Four marbles are in the bottom row. **Subtract** four from six to find how many more marbles are on the top row.

See also: **minus, subtraction**

subtraction (−)
/suhb TRAK shuhn/ (n.)

the operation that is the inverse of addition

$$15 − 5 = 10$$

$$4 = 7 − 3$$

$$\frac{1}{2} − \frac{1}{4} = \frac{1}{4}$$

Use **subtraction** to find the difference between two numbers.

See also: **addition, inverse, minus, subtract**

sum /suhm/ (n.)

the answer to an
addition problem

27 + 7 = 34

The **sum** of 27 and 7 is 34.

See also: **addition, difference, product, quotient**

supplementary angle
/suh pluh MEN tuh ree ANG guhl/ (n.)

one of two angles whose sum
is 180°

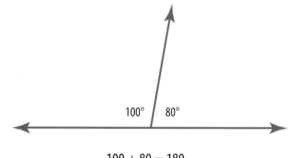

100 + 80 = 180

See also: **angle, complementary angle, measure, straight angle, sum**

Daniel draws a pair of **supplementary angles**.
One angle measures 100°. The other measures 80°.

surface /SUR fuhs/ (n.)

the outside of a solid figure

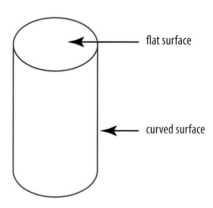

flat surface

curved surface

See also: **curve, face, flat, solid**

Cylinders have one curved **surface**
and two flat **surfaces**.

symbol   /SIM buhl/ (n.)

a mark or sign that stands
for words or numbers

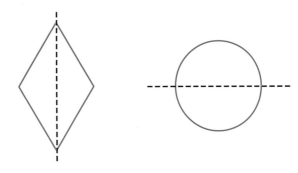

CCVI   π   ÷   <   ≈   %   °C

Each **symbol** has a certain meaning.
The **symbol** "÷" means "divide."

See also: **sign**

symmetrical

/si MET ruh kuhl/ (adj.)

divisible into two
mirror images

See also: **symmetry**

Use a line of symmetry to show that an object
is **symmetrical**.

symmetry   /SI muh tree/ (n.)

the property of having two
halves that are mirror images

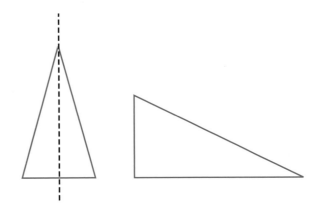

See also: **line of symmetry,
rotational symmetry, symmetrical**

An isosceles triangle has **symmetry** called
line symmetry. A scalene triangle does not.

Tt

table /TAY buhl/ (n.)
a set of data shown
in an organized way

×	1	2	3	4	5
1	1	2	3	4	5
2	2	4	6	8	10
3	3	6	9	12	15
4	4	8	12	16	20
5	5	10	15	20	25

See also: **chart, column, data, row**

This **table** shows some of the multiplication facts.
There are six rows and six columns.

---

take away /tayk uh WAY/ (v.)
to subtract

See also: **add, minus, subtract**

Eight **take away** four is four.

---

tally /TA lee/ (n.)
a mark to represent a number
or quantity

See also: **count, tally chart**

A **tally** shows one. A line through four **tallies**
shows five.

tally chart /TA lee chart/ (n.)

a display of data that shows
frequency with tallies

Grade	Number of Students, or Frequency
90–100	IIII
80–89	HHt HHt II
70–79	HHt III
60–69	II

See also: **count, frequency, table, tally**

This **tally chart** shows grades on a test.
Twelve students scored between 80 and 89.

tangram /TAN gruhm/ (n.)

a puzzle of seven pieces
in a square shape

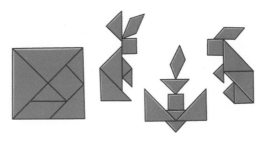

See also: **parallelogram, square, triangle**

There are five triangles in a **tangram**.
There are also a parallelogram and a square.

tape measure
/tayp MEZH ur/ (n.)

a long, thin tape or ribbon
for measuring length

See also: **centimeter, inch, length, ruler**

Use a **tape measure** to measure the size of a room.

temperature
/TEM pur uh chur/ (n.)

the measure or degree of heat

Water usually freezes at a **temperature**
of 0°C (32°F).

See also: **Celsius, degree, Fahrenheit,
thermometer**

135

tessellation
/te suh LAY shuhn/ (n.)

a pattern of shapes without gaps or overlaps

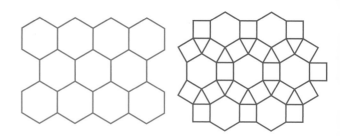

See also: **pattern**

You can make a **tessellation** from one shape or from many shapes.

tetrahedron
/te truh HEE druhn/ (n.)

a solid with four flat faces that are triangles

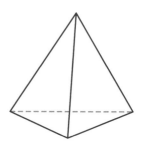

See also: **face, pyramid, solid, vertex**

The **tetrahedron** has four vertices, or corners.
A **tetrahedron** is also called a triangular pyramid.

thermometer
/thur MOM uh tur/ (n.)

a tool for measuring temperature

See also: **Celsius, degree, Fahrenheit, temperature**

This **thermometer** shows the temperature. It is about 25°C, or 77°F.

third /thurd/ (adj.)

after the first and second in a sequence

15, 30, 45, 60, 75, 90, ...

The **third** number in the sequence is 45.

See also: **ordinal, second, sequence**

three-dimensional (3-D)
/three duh MEN shuhn uhl/ (adj.)

having points or sides that are
not all on one plane

See also: **dimension, plane, solid,
two-dimensional**

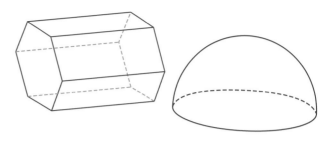

Each **three-dimensional** figure has length, width,
and height.

time /tīm/ (v.)

to keep track of seconds,
minutes, or hours

See also: **hour, minute, second**

Time how long it takes you to run a mile.
Use a watch to **time** your run.

time /tīm/ (n.)

1. how long something lasts

Measure **time** in seconds, minutes, hours,
days, weeks, months, and years.

2. a point in the day shown as
   an hour and a certain number
   of minutes

See also: **century, hour, minute, second**

The **time** is 7:00 p.m.

timer /tī mur/ (n.)

a tool for measuring time

See also: **hour, minute, second, time**

Measure seconds, minutes, or hours with a **timer**.
A stopwatch is a kind of **timer**.

137

times (×) /tīmz/ (prep.)
multiplied by

$$4 \times 5 = 20$$

Four **times** five equals twenty.

See also: **multiply**

times table /tīmz TAY buhl/ (n.)
an organized display of
the multiplication facts

×	1	2	3	4	5	6
1	1	2	3	4	5	6
2	2	4	6	8	10	12
3	3	6	9	12	15	18
4	4	8	12	16	20	24
5	5	10	15	20	25	30
6	6	12	18	24	30	36

Use the **times table** to find 4 times 5. Go to 4 in the fifth row. Go to 5 in the sixth column. The row and column meet at 20. $4 \times 5 = 20$

See also: **multiple, multiplication, table, times**

ton /tuhn/ (n.)
a unit of weight in the
customary system

1 **ton** = 2,000 pounds

See also: **customary system, pound, weight**

The weight of an average small car is about two **tons**.

total /TOH tuhl/ (adj.)
made up of the whole amount or number

We spent $100 on juice and $200 on sandwiches. The **total** cost of the school picnic was $300.

See also: **add, cost, sum**

total /TOH tuhl/ (v.)
to find the full amount or number

I scored two points. Sara scored four points. Ivan scored three points.
We **total** the points to find our team's score:
2 + 4 + 3 = 9 points

See also: **add, sum**

trace /trays/ (v.)
to copy a figure by drawing a line on see-through paper

Trace a triangle on a piece of see-through paper.

Use this copy to help you understand transformations.

See also: **reflection, rotation, transformation, translation**

transformation
/tran sfur MAY shuhn/ (n.)
to move a figure in a particular way

See also: **figure, reflection, rotation, translation**

Three common **transformations** are rotations, translations, and reflections. This picture shows a reflection of figure A.

translate /tranz LAYT/ (v.)

to slide a figure on a straight line to a new location

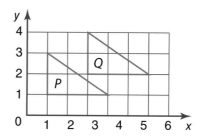

See also: **distance, slide, transformation, translation**

Translate triangle *P*. Move every point the same distance. The result is triangle *Q*.

translation /tranz LAY shuhn/ (n.)

a change of a figure's position to a new location

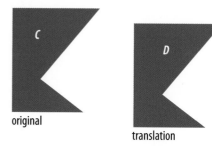

original

translation

See also: **distance, slide, transformation, translate**

Every point on figure *C* was moved the same distance. Figure *C* did not change in size. Figure *C* did not rotate. Figure *D* is a **translation** of figure *C*.

trapezoid /TRA puh zoid/ (n.)

a quadrilateral with exactly one pair of parallel sides

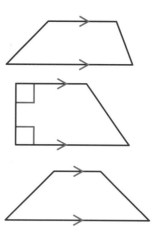

See also: **parallel, parallelogram, quadrilateral, right angle**

One pair of sides is parallel in a **trapezoid**. The figure is a parallelogram if both pairs are parallel.

tree diagram
/tree DĪ uh gram/ (n.)

a picture with branching lines
that models information

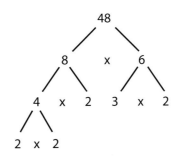

See also: **diagram, factor**

A **tree diagram** shows the factors of a number.
This **tree diagram** shows the factors of 48.

trial /TRĪ uhl/ (n.)

one part of an experiment
to test probability

See also: **event, number cubes,
probability, random**

Roll a number cube ten times. Each roll is one **trial**.
Repeat the **trials** to check the probability of rolling a 6.

trial and error
/TRĪ uhl and ER ur/ (n.)

a process of making
reasonable guesses
to solve a problem

You can use **trial and error** to find x in 14 + x = 22.
Your first guess is 7, but 14 + 7 = 21.

Your second guess must be more than 7.
You guess 9, but 14 + 9 = 23.

The answer must be between 7 and 9: 14 + 8 = 22.

See also: **guess, solve**

triangle /TRĪ ang guhl/ (n.)

a polygon with three sides

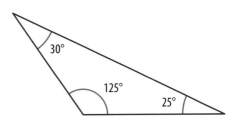

See also: **angle, degree, polygon, side**

A **triangle** has three angles. The sum of
the measures of the angles is always 180°.

triangle number

/TRĪ ang guhl NUHM bur/ (n.)

a number that can be shown with
a particular triangle pattern

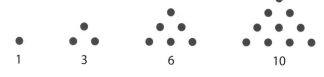

1   3   6   10

Each **triangle number** is shown by dots arranged
as an equilateral triangle.

See also: **equilateral triangle, pattern,
square number**

triangular pyramid

/trī ANG gyuh lur PEER uh mid/ (n.)

a pyramid with a base
that is a triangle

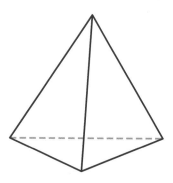

See also: **base, pyramid,
tetrahedron, vertex**

The **triangular pyramid** has four vertices, or corners.

triple   /TRI puhl/ (adj.)

three times as much or as many

A 300-degree angle has **triple** the measure
of a 100-degree angle.

See also: **angle, degree, double, multiply**

two-dimensional (2-D)

/too duh MEN shuhn uhl/ (adj.)

having length and width
but no depth or height

All of the points of each **two-dimensional** figure
lie in one plane.

See also: **dimension, plane, solid,
three-dimensional**

unequal (≠) /uhn EE kwuhl/ (adj.)

not the same in size, degree, or amount

25 ≠ 8 × 3

Twenty-five is not equal to eight times three. They are **unequal**.

See also: **equal, times**

union (∪) /YOON yuhn/ (n.)

the set of all elements in two or more sets

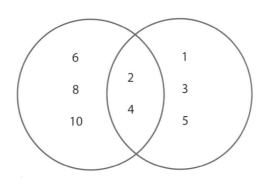

Set A = {2, 4, 6, 8, 10}
Set B = {1, 2, 3, 4, 5}

The **union** of sets A and B (A ∪ B) is the set with elements in either A or B.

See also: **intersection, set, Venn diagram**

unit /YOO nuht/ (n.)

a quantity that is the standard when measuring something

A liter is a **unit** for measuring capacity.
A pound is a **unit** for measuring weight.
Inches and meters are **units** for measuring length.

See also: **cubic unit, measurement, square unit**

Vv

V /vee/ (n.)

the Roman numeral
representing 5

$$V = 5$$
$$VI = 6$$
$$CV = 105$$

See also: **C, I, Roman numerals, X**

---

value /VAL yoo/ (n.)

1. the amount of money
that something is worth

The car's **value** is about $5,000,
but I paid only $4,000.

2. the worth of a variable
in an equation

Look at this equation: $x + y = 45$
If the **value** of x is 15, the **value** of y is 30.

3. the worth of a fraction

The fractions $\frac{1}{5}$ and $\frac{2}{10}$ are equivalent.
Each has the same **value**.

4. the worth of a number's digit

hundreds   tens   ones

173

The **value** of the 1 is 100.
The **value** of the 7 is 70.

See also: **digit, equation, equivalent
fraction, variable**

variable /VAIR ee uh buhl/ (n.)
a letter or symbol that represents a number

> A **variable** can represent one number. It can also represent many numbers or all numbers.
>
> Here the **variable** m stands for one number:
> $m + 12 = 19$

See also: **algebra, equation, represent, symbol**

variable /VAIR ee uh buhl/ (adj.)
able to have different values

> I made a graph of the river's depth over time. The depth is **variable**. It changes over time.

See also: **depth, graph, time, value**

Venn diagram
/ven DĪ uh gram/ (n.)

a picture with circles that shows sets and their relationship

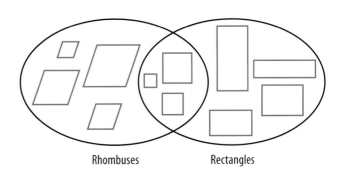

Rhombuses   Rectangles

See also: **rectangle, rhombus, set, square**

The **Venn diagram** shows rhombuses and rectangles. Squares are both rhombuses and rectangles.

vertex (plural vertices)
/VUR teks/ (n.)

the point where sides
or edges meet

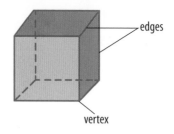

edges

vertex

See also: **corner, cube, edge, side**

A cube has eight **vertices**. A square has
four **vertices**, or corners.

vertical /VUR ti kuhl/ (adj.)

perpendicular to the horizon;
going straight up and down

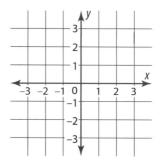

See also: **horizontal, line, perpendicular,
right angle**

The y-axis is a **vertical** line. It is at a right angle
to the x-axis.

volume /VOL yuhm/ (n.)

1. the amount a solid holds

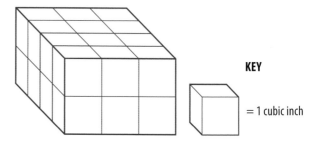

KEY

= 1 cubic inch

Volume is measured in cubic units. The **volume**
of the rectangular prism is 24 cubic inches.

2. the amount of something
measured in a container

See also: **capacity, cubic unit, gallon**

A container holds 1 gallon of milk.
The **volume** of the milk is 1 gallon.
The capacity of the container is 1 gallon.

Ww

weigh /way/ (v.)

to measure how heavy
an object is

See also: **balance, heavy, measure, scale**

Use a scale or balance to **weigh** an object.
Weigh the object using ounces, pounds, or tons.

weight /wayt/ (n.)

the measure of how heavy
an object is

Weight is the force of gravity on an object.
It is the force that pulls an object toward
the center of the Earth.

See also: **heavy, mass, measure**

whole /hohl/ (n.)

all of an object, a collection
of objects, or a quantity

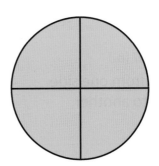

See also: **circle, fraction, part**

The circle is divided into four equal parts.
You can represent the **whole** as 1 or as the fraction $\frac{4}{4}$.

147

whole number
/hohl NUHM bur/ (n.)

one of the numbers for counting, including zero

0, 1, 2, 3, 4, ...

A **whole number** is a counting number.
A fraction is not a **whole number**.

See also: **integer, positive number, rational number, zero**

---

wide /wīd/ (adj.)

from one side of an object to another

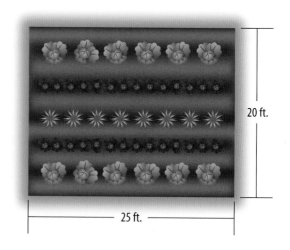

20 ft.

25 ft.

See also: **length, long, measure, width**

Mr. Anders measures the garden.
It measures 25 feet long by 20 feet **wide**.

---

width /width/ (n.)

the distance from one side of an object to another

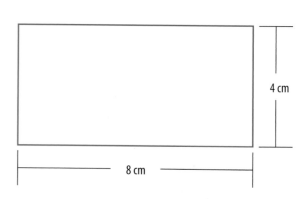

4 cm

8 cm

See also: **distance, length, measure, rectangle**

A rectangle has length and **width**.
The rectangle's **width** is 4 centimeters.

X /eks/ (n.)

the Roman numeral
representing 10

X = 10
XI = 11
C**X** = 110

See also: **C, I, Roman numerals, V**

X /eks/ (n.)

a letter commonly used to name
a variable

4*x* = 20

What is the value of *x* in this equation?
The value of *x* is 5.

See also: **algebra, equation, value, variable**

---

x-axis /eks AK suhs/ (n.)

the horizontal line in
a coordinate grid

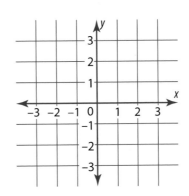

See also: **coordinate grid, cross, horizontal, *y*-axis**

The ***x*-axis** is a horizontal line. The *y*-axis is a vertical line. The ***x*-axis** crosses the y-axis at (0, 0).

x-coordinate
/eks koh ORD nit/ (n.)

the first number in an ordered pair; the horizontal position of a point

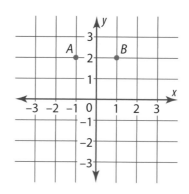

See also: **coordinate grid, horizontal, ordered pair, point, y-coordinate**

The **x-coordinate** of point *A* is −1. The **x-coordinate** of point *B* is 1.

Yy

yard /yard/ (n.)
a unit of length in the customary system

1 **yard** = 3 feet
1 mile = 1,760 **yards**

See also: **customary system, foot, inch, mile**

y-axis /wī AK suhs/ (n.)
the vertical line of a coordinate grid

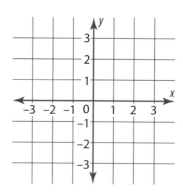

See also: **coordinate grid, perpendicular, vertical, x-axis**

The **y-axis** is a vertical line. The **y-axis** is perpendicular to the *x*-axis.

y-coordinate
/wī koh ORD nit/ (n.)

the second number in an
ordered pair; the vertical position
of a point

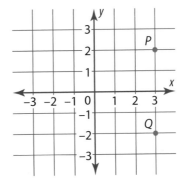

See also: **coordinate grid, ordered pair,
point, vertical, x-coordinate**

The **y-coordinate** of point *P* is 2. The **y-coordinate**
of point *Q* is −2.

zero (0)   /ZEE roh/ (n.)

1. a number that is
   not positive or negative

   The number 0, or **zero**, is between −1 and 1.
   … −3, −2, −1, 0, 1, 2, 3 …

2. a temperature on the
   Fahrenheit or Celsius scale

   Water freezes at **zero** degrees Celsius (0°C).
   Zero degrees Celsius is the same as 32
   degrees Fahrenheit (32°F).

3. no quantity; nothing

   The cup is empty. There are **zero** ounces
   in the cup.

See also: **degree, integer,
negative number, ounce**

Number and Operations

abacus
add
addition
altogether
answer
approximate
approximately
arithmetic
array
ascending
Associative Property
backward
base
before
billion
brackets
button
C
calculate
calculation
calculator
change
cheap
check
coin
common denominator
Commutative Property
compare
consecutive
correct
cost
count
counter
cube
cube root
D
decimal

decimal place
decimal point
decrease
denominator
descending
difference
digit
discount
Distributive Property
divide
dividend
divisible
division
divisor
double
empty
element
equal
equals sign
equivalent fraction
Eratosthenes sieve
estimate
estimation
even
exact
exchange
expanded form
exponent
factor
flat
forward
fraction
greater than
greatest common factor
guess
half
I

improper fraction
increase
inequality
infinity
integer
interest
interval
inverse
L
least common multiple
less than
long
M
million
minus
mixed number
multiple
multiplication
multiply
negative number
number
number fact
number line
numeral
numerator
odd
operation
opposite
order
order of operations
ordinal
parentheses
pattern
pay
per
percent
perfect square

place
place value
plus
positive number
power
price
prime factor
prime number
product
profit
proper fraction
proportion
quarter
quotient
rate
ratio
rational number
reduce
regroup
remainder
repeat

repeating decimal
represent
Roman numerals
second
sell
sequence
set
share
sign
simplify
solution
solve
spend
square
square number
square root
squared
standard form
step
subset
subtract

subtraction
sum
symbol
take away
third
times
times table
total
tree diagram
trial and error
triangle numbers
triple
unequal
union
V
value
whole
whole number
X
zero

Expressions, Equations, and Algebra

algebra
dependent variable
direct proportion
equal
equals sign
equation
exponent
expression
formula
greater than

independent variable
inequality
input values
inverse
inverse proportion
mapping
number sentence
output values
parentheses
pattern

rate
represent
rule
sign
symbol
unequal
value
variable
x

Geometry

acute angle	convex	face
acute triangle	coordinate grid	figure
adjacent	coordinates	find
altitude	copy	flat
angle	corner	flip
apex	counterclockwise	fold
arc	cover	formula
area	cross	geoboard
attribute	cross-section	geometry
axis	cube	graph
axis of rotation	cubic unit	group
balance	curve	half
base	customary system	hemisphere
bisect	cylinder	heptagon
big	D	hexagon
center	decagon	hexagonal prism
center of rotation	decahedron	hexahedron
centiliter	degree	horizontal
centimeter	depth	icosahedron
clockwise	diagonal	identical
chord	diagram	intersect
circle	diameter	intersection
circular	diamond	irregular
circumference	different	irregular polygon
compare	dimension	isosceles triangle
compass	distance	kite
compass points	dodecagon	label
complementary angle	dodecahedron	line
concave	edge	line of symmetry
concentric	ellipse	line segment
cone	end point	map
congruent	enlarge	median
construct	equilateral triangle	mid-
convert	equivalent	middle

narrow
net
nonagon
oblique
oblong
obtuse angle
obtuse triangle
octagon
octahedron
opposite
ordered pair
origin
oval
pair
parallel
parallelogram
part
pattern
pentagon
pentomino
perimeter
perpendicular
pi
plane
plane of symmetry
plane shape
plot
point
polygon
polyhedron
position
prism
property

proportion
pyramid
quadrant
quadrilateral
radius
ray
rectangle
reduce
reflect
reflection
reflective symmetry
reflex angle
regular polygon
regular polyhedron
revolve
rhombus
right angle
right triangle
rotate
rotation
rotational symmetry
round
scale
scale drawing
scalene triangle
sector
semicircle
shade
short
side
size
slide
solid

sphere
spherical
spiral
square
square unit
straight angle
supplementary angle
surface
symmetrical
symmetry
tangram
tessellation
tetrahedron
third
three-dimensional
trace
transformation
translate
translation
trapezoid
triangle
triangular pyramid
two-dimensional
unit
vertex
vertical
volume
whole
x-axis
x-coordinate
y-axis
y-coordinate

Measurement and Time

accurate	gram	morning
after	heavy	night
afternoon	height	ounce
a.m.	hour	pint
area	hour hand	p.m.
before	inch	pound
calendar	interval	quart
capacity	kilogram	ruler
Celsius	kilometer	scale
century	leap year	second
clock	length	short
clockwise	liter	slow
counterclockwise	long	speed
cubic unit	mass	square unit
cup	measure	standard unit
D	measurement	tape measure
date	meter	temperature
day	meter stick	thermometer
deciliter	metric system	time
decimeter	metric ton	timer
degree	midnight	ton
digital	mile	unit
dimension	millennium	volume
empty	milliliter	weigh
Fahrenheit	millimeter	weight
foot	minute	wide
full	minute hand	width
gallon	month	yard

Data and Probability

arrange
average
bar graph
certain
chart
choose
circle graph
column
curve
data
database
event
flip
frequency

impossible
label
likely
line plot
maximum
mean
median
minimum
mode
number cubes
outcome
pictogram
pie chart
predict

probability
probable
questionnaire
random
range
record
row
table
tally
tally chart
tree diagram
trial

Modeling and Problem Solving

answer
approximate
array
calculator
check
coordinate grid
cube
diagram
flat
flow chart

guess
interlocking cubes
investigate
list
long
model
net
number line
pattern
reasonable

represent
scale drawing
solution
solve
step
tree diagram
trial and error
Venn diagram

Mathematical Symbols

Symbol	Meaning	Example
| |	absolute value	$\|-3\|$
$\angle$	angle	$\angle QRS$
$\approx$	approximately equal to	$\sqrt{3} \approx 1.732$
¢	cents	37¢
$\cong$	congruent to	$\triangle ABC \cong \triangle DEF$
$\circ$	degrees	12°C
$\div$	divide	$8 \div 4 = 2$
/	divide	$8/4 = 2$
$\overline{)}$	divide	$8\overline{)64}^{\,8}$
$\neq$	does not equal	$2 + 3 \neq 1 + 5$
$	dollars	$23.00
=	equals	$8/4 = 2$
!	factorial	$3! = 3 \times 2 \times 1$
( )	grouping	$(6 + 6) - (3 + 5)$
∞	infinity	$1, 2, 3, \ldots \infty$
$\cap$	intersection	$\{2, 3\} \cap \{3, 4\} = \{3\}$
<	less than	$4 < 7$
$\leq$	less than or equal to	$1 \leq 4$
>	greater than	$1 > -3$
$\geq$	greater than or equal to	$-4 \geq -4$

Symbol	Meaning	Example
$\times$	multiply	$4 \times 8 = 32$
$\cdot$	multiply	$4 \cdot 8 = 32$
$*$	multiply	$4 * 8 = 32$
( )	multiply	$4(8) = 32$
—	negative	-6
$\|\|$	parallel	$\overline{AB} \parallel \overline{CD}$
%	percent	100%
$\perp$	perpendicular	$AB \perp CD$
π	pi	$\pi \approx 3.1416$
+	plus	$2 + 2 = 4$
$\pm$	plus or minus	$2a = \pm 3$
+	positive	$+5$
:	ratio	$3 : 2$
—	repeating decimal	$4.\overline{3}$
$\sim$	similar to	$\triangle ABC \sim \triangle DEF$
$\sqrt{}$	square root	$\sqrt{16}$
—	subtract	$31 - 8 = 23$
$\triangle$	triangle	$\triangle ABC$
$\cup$	union	$\{2, 3\} \cup \{3, 4\} = \{2, 3, 4\}$

Formulas for Perimeter

Polygon	Formula	Diagram
square	$P = 4s$	
rectangle	$P = 2l + 2w$	
triangle	$P = a + b + c$	
parallelogram	$P = 2a + 2b$	
circle (circumference)	$C = 2\pi r$	

Formulas for Area

Polygon	Formula	Diagram
square	$A = s^2$	
rectangle	$A = lw$	
triangle	$A = \frac{1}{2}bh$	
parallelogram	$A = bh$	
circle	$A = \pi r^2$	

Formulas for Volume

Solid	Formula	Diagram
cube	$V = s^3$	
rectangular prism	$V = lwh$	
cone	$V = \frac{1}{3}\pi r^2 h$	
cylinder	$V = \pi r^2 h$	

Multiplication Table

×	1	2	3	4	5	6	7	8	9
1	1	2	3	4	5	6	7	8	9
2	2	4	6	8	10	12	14	16	18
3	3	6	9	12	15	18	21	24	27
4	4	8	12	16	20	24	28	32	36
5	5	10	15	20	25	30	35	40	45
6	6	12	18	24	30	36	42	48	54
7	7	14	21	28	35	42	49	56	63
8	8	16	24	32	40	48	56	64	72
9	9	18	27	36	45	54	63	72	81

Place-Value Chart

Millions	Hundred Thousands	Ten Thousands	Thousands	Hundreds	Tens	Ones	.	Tenths	Hundredths	Thousandths
5	7	1	3	5	2	8	.	6	4	7

Important Math Properties

Property	Example
Associative Property of Addition	$(1 + 2) + 3 = 1 + (2 + 3)$
Associative Property of Multiplication	$(2 \times 3) \times 4 = 2 \times (3 \times 4)$
Commutative Property of Addition	$1 + 2 = 2 + 1$
Commutative Property of Multiplication	$2 \times 4 = 4 \times 2$
Distributive Property	$2 \times (3 + 4) = (2 \times 3) + (2 \times 4)$
Identity Property of Addition	$1 + 0 = 1$
Identity Property of Multiplication	$2 \times 1 = 2$
Multiplication Property of Zero	$3 \times 0 = 0$

Converting Basic Metric Units to Customary Units

Metric Unit	Conversion Factor
centimeter	× 0.394 = inches
meter	× 3.281 = feet
kilometer	× 0.621 = miles
gram	× 0.035 = ounces
kilogram	× 2.205 = pounds
degrees Celsius	(°C × 1.80) + 32 = degrees Fahrenheit
milliliter (cubic centimeter)	× 0.034 = fluid ounces
liter	× 0.264 = gallons
joule	× 0.239 = calories

Converting Basic Customary Units to Metric Units

Customary Unit	Conversion Factor
inch	× 2.540 = centimeters
foot	× 0.305 = meters
mile	× 1.609 = kilometers
ounce	× 28.350 = grams
pound	× 0.454 = kilograms
degrees Fahrenheit	(°F − 32) × 0.556 = degrees Celsius
fluid ounce	× 29.574 = milliliters (cubic centimeters)
gallon	× 3.785 = liters
calorie	× 4.184 = joules

Measurements of Time

60 seconds = 1 minute
60 minutes = 1 hour
24 hours = 1 day
28, 29, 30, or 31 days = 1 month
365 or 366 days = 1 year
10 years = 1 decade
100 years = 1 century
1,000 years = 1 millennium

Metric Prefixes

Prefix	Meaning	Length	Mass	Volume
kilo-	1,000	kilometer (km)	kilogram (kg)	kiloliter (kl)
hecto-	100	hectometer (hm)	hectogram (hg)	hectoliter (hl)
deka-	10	dekameter (dam)	dekagram (dag)	dekaliter (dal)
(none)	1	meter (m)	gram (g)	liter (l)
deci-	0.1	decimeter (dm)	decigram (dg)	deciliter (dl)
centi-	0.01	centimeter (cm)	centigram (cg)	centiliter (cl)
milli-	0.001	millimeter (mm)	milligram (mg)	milliliter (ml)

Squares and Cubes

Number	Squared (n^2)	Cubed (n^3)
1	1	1
2	4	8
3	9	27
4	16	64
5	25	125
6	36	216
7	49	343
8	64	512
9	81	729
10	100	1,000
11	121	1,331
12	144	1,728
13	169	2,197
14	196	2,744
15	225	3,375
16	256	4,096
17	289	4,913
18	324	5,832
19	361	6,859
20	400	8,000